NUTRITION AND BRAIN DEVELOPMENT

NUTRITION AND BRAIN DEVELOPMENT

Govind A. Dhopeshwarkar

Late of University of California at Los Angeles
School of Public Health
Los Angeles, California

PLENUM PRESS • NEW YORK AND LONDON

Library of Congress Cataloging in Publication Data

Dhopeshwarkar, Govind A., 1924–1982.
    Nutrition and brain development.

    Bibliography: p.
    Includes index.
    1. Brain—Diseases—Nutritional aspects. 2. Brain—Growth. 3. Malnutrition—Complications and sequelae. I. Title. [DNLM: 1. Brain—Growth and development. 2. Nutrition. 3. Nutrition disorders—Complications. WL 300 D535n]
RC386.2.D448 1983   616.8'0471   83-8139
ISBN 0-306-41060-5

Following the death of the author, the editorial production of this volume was supervised by James F. Mead, Professor of Biochemistry, University of California at Los Angeles, Los Angeles, California

©1983 Plenum Press, New York
A Division of Plenum Publishing Corporation
233 Spring Street, New York, N.Y. 10013

Printed in the United States of America

IN MEMORIAM
Govind Atmaram Dhopeshwarkar
1923–1982

With the complete draft of this book already submitted, Dr. Dhopeshwarkar died April 1, 1982, after being stricken by a heart attack while playing his favorite game of tennis. He is survived by his wife, Saral, and two daughters, Anita Dhopeshwarkar and Rohini Joshi. He was Research Biochemist, Laboratory of Biomedical and Environmental Sciences, and Adjunct Professor, Environmental and Nutritional Sciences, School of Public Health, University of California at Los Angeles.

Nana, as he was known by his many friends, was a native of India and was educted in Bombay, receiving a Ph.D. in Biochemistry from the University of Bombay in 1954. After postdoctoral positions in India, Sweden and America, he went to UCLA in 1967 where he focused his research on nutrition in the developing brain including the biosynthesis, interconversions, and turnover of essential and nonessential unsaturated fatty acids. He was also an investigator of the effects of dietary egg cholesterol on blood lipids of young and middle-aged men.

In addition to his scientific achievements, Nana was universally loved by his colleagues and employees. He was characterized by such gentleness that he was never heard to say an unkind word about anyone, no matter what the provocation. His many friends and associates will miss him.

James F. Mead

PREFACE

The effect of nutrition on the vital process of brain development has received increased attention in the last two decades. Using animal models, experimental research scientists have accumulated a wealth of information and epidemiological studies of field workers have brought the animals and human studies together. Most times, there is an agreement on the results of animal and human experiences, but occasionally a voice of uncertainty is heard when results of animal studies are extrapolated to humans. After all, the human brain is far more complex than that of other species, and comparisons are not always accurate. Behavioral scientists have attempted to correlate the findings of the biochemists and neurochemists to the "working" of the brain.

Severe effects of malnutrition on body growth and function can usually be reversed by corrective procedures. But when such effects include the impact on brain development, the outcome can be devastating. Underdevelopment of the brain caused by malnutrition during early life may not respond to corrective measures in later life. Undoubtedly this is a very controversial issue and the final verdict has not been reached. Unfortunately, even with today's tremendous technological innovations and applications, there are millions of people, including infants and children, who cannot obtain adequate food and are therefore exposed to damaging effects on the orderly development of the central nervous system. Thus, this is a critical issue to many struggling populations who can ill afford such deprivation.

This book is offered as a continuous woven fabric to encompass all major aspects of brain development and the effects of nutrition on this process. I have started with a brief description of the morphology and cellular makeup of the brain to understand the working of the brain. The most important aspects of brain development are related to timing of the rapid growth. During this relatively short period of growth, referred to as the critical period or growth spurt, the brain seems to be more vulnerable to effects of malnutrition and environmental insults. This explains why the effects of malnutrition on the brain have been separated from the effects on the rest of the body. We continue to describe the effects of not just caloric deficiency but include the effects of protein and the essential fatty acid deficiencies. Some aspects of vitamin deficiencies and exposure to toxic compounds have also been included. The blood–brain barrier system is discussed since it certainly mediates the availability of the nutrient supply to the brain. The energy requirements of the brain are discussed with added new concepts. Emphasis is placed on the fact that maternal nutrition, seemingly adequate in calories but imbalanced in required components or sometimes overly rich in certain factors, can be damaging to the proper development of the brain.

Recent findings related to nutrition and regulation of neurotransmitter levels and discoveries of neuropeptides that have excited neurochemists and neurophysiologists have been briefly explained. Finally, some inborn errors of metabolism that affect the central nervous system have been included and recent advances toward the attempt to overcome these have been included.

In summary, this book covers a wide variety of related subjects. The special emphasis on the effects of an imbalanced diet, as against lack of food, should be of value to people living in countries that are relatively prosperous and have successfully overcome hunger. The complexities of the central nervous system and the difficulty in assigning cause and effect during the long life span of humans makes the subject of nutrition and human brain development very difficult to explain. But this is the challenge of the times and the most gifted brains are constantly working to seek answers.

This book could not have been written without the aid of colleagues and mentors. Dr. James F. Mead, Professor of Biological Chemistry, and Roslyn B. Alfin-Slater, Professor of Nutrition at UCLA, have been a source of inspiration for me for several years, and I owe them my deepfelt gratitude for the encouragement they have given in my research and teaching. Much of the work on brain lipid biochemistry cited in the book was a collaborative effort and included valuable input from several colleagues and students. I would like to mention the following who have made very valuable contributions: Carole Subramanian, Rona Karney, Carolyn Moore, Susan Bony, Nirmala Menon, Maria Martins, and Annette Aftergood. Finally, without the help of Joyce Adler, J. Yamaguchi, and Margaret Bachtold who spent endless hours in typing and Dr. Lilla Aftergood, who offered very valuable advice in editing, it would not have been possible to put the manuscript in its final form. I sincerely thank all those who made this possible.

G. A. Dhopeshwarkar

CONTENTS

4. Blood–Brain Barrier System

5. Alcohol: Effects on the Central Nervous System

6. Effects of Malnutrition on Brain Development

7. Vitamin Deficiencies and Excesses

INTRODUCTION

Of all the mysteries of life, the working of the brain still remains the main challenge. We know many things about how the brain functions, yet when interpreting higher functions of the brain we admit we know very little. Even with all the advances in methodology, and many have been spectacular, and the ever present computer, we still are unable to fully understand and explain the higher functions of the brain. Other organs of the body, although very important to the preservation of life, have a defined function: the heart is a pump, the lungs are an oxygenating system, and an endocrine gland is entrusted with the task of synthesis and secretion of single or multiple hormones (e.g., thyroid, pituitary). The brain, however, stands out all by itself: it is the master control, and, although its control of physical acts may be understood, we are still unable to explain in biochemical or physiological terms (which we can understand today) how the brain controls and creates thought, behavior, mood, and emotions. This may be the greatest challenge to human efforts and the reason why the most brilliant minds that were busy solving the problems of genetics and inheritance are now excited by the mysteries of brain functions and are actively engaged in research in this field.

The size of the brain and its capability to perform complex functions do not parallel one another. For example, the brains of elephants and whales are many times larger than man's but are incapable of performing the tasks that man can perform easily. Curiously enough, the brain of a dolphin, which is much smaller in size than that of a whale or elephant, is slightly larger than that of man. In this connection the ratio of body weight to brain weight may be more valuable; for example, in the adult human, the ratio is approximately 50, whereas in the rat it is about 200 and in a monkey it is about 80.

The complexities of the brain are enormous. A casual look at the anatomy of the brain shows us that there are distinct areas of the brain such as cerebral cortex, cerebellum, medulla and hippocampus, in addition to the so-called "master gland," the pituitary, situated in the brain area. Each area has been associated with some major functions. For example, the multilobed cerebrum regulates speech, vision, and hearing. The amygdala, hippocampus, and hypothalamus, parts of the limbic system, control emotions. The pituitary produces and secretes many hormones and release factors that regulate diverse functions including growth and development. Although we have been able to map the many events that occur during this growth and development, we are still far from understanding clearly how these correlate with maturity of thought. By first taking apart different areas of the brain and using various *in vitro* enzymatic methods in order to simplify the complex system, one can learn how the growth of the brain affects its biochemical makeup. Undoubtedly, using this approach we have discovered many metabolic reactions: how components are synthesized and degraded, and, in a physiological sense, how the signals originate and are conducted from one neuron to the next. Another approach has been to use cells in culture. Although it is difficult to maintain normal neuronal cells continuously in culture, such methods have recently been developed for glial cells. Experiments in pure cell lines are extremely valuable and productive. It may be pointed out here that due to a unique phenomenon, the blood–brain barrier, any process that bypasses this, like *in vivo* enzymatic studies or studies with cell culture, always leaves a doubt: Can the *in vitro* methods be equated to *in vivo* conditions? Even when the *in vivo* approach is used, the method of introduction of the compound, appropriate dose, and effect of anesthesia have been questioned. At this time, all the approaches are needed to arrive at an understanding of the complex working of the brain.

We are concerned with growth and development and would like to focus on the effects of nutrition on this important phase of brain development. The impact of nutritional stress could come via several angles. For example, one could start with caloric deficiency, common to many areas around the world, deficiency of one or multiple major components, e.g., proteins, fats, or deficiency of single or multiple vitamins or minerals. It has been proven, mostly by animal experiments, that lack of essential amino acids or essential fatty acids is deleterious to brain development. In other words, nutritional imbalance influences the growth and development of the brain; but it does not have to be restricted to deficiency alone, excesses in dietary intake are also harmful. Take the example of ingestion of alcohol, caffeine, or nicotine; all have been proven to result in a harmful effect of one kind or the other.

Environmental stresses like heavy-metal poisoning arising from careless industrial policies, atmospheric pollution, and poor water quality all take their toll. Many of the effects could be subtle and only careful analysis could sort them out.

A very important additional factor needs special attention. The growth characteristics of the brain are unique and differ from the rest of the body in terms of timing. One can easily infer that the maximum impact of nutritional imbalances, deficiencies as well as overabundances, would be felt during the period of maximum growth. In general, the period of maximum growth in the brain is over long before that of the rest of the body and is species dependent. This vulnerable or critical period is unfortunately not uniform throughout the brain, i.e., some parts of the brain start the accelerated growth a little before other parts and may continue longer. Complicated time-schedule experiments with various types of diet (representing lack or overabundance of nutrients) to look at various regions have not been performed in all cases. However, effects of various nutrients on different cell types that make up the various regions of the brain have been carried out to arrive at some general conclusions. Thus, the complicated situation does not look desperate! One can even say that a significant inroad has been made to assess the effects of nutritional imbalances on the central nervous system.

To understand these effects, therefore, it is necessary to know the details of growth characteristics and cellular makeup of the brain and the biochemical reactions taking place in the central nervous system.

1

CELLULAR MAKEUP OF THE BRAIN

Without going into detailed and complicated anatomy of the central nervous system and the various characteristics of specialized cells that make up different areas of the brain, one can broadly categorize the cellular makeup of the central nervous system.

1.1. NEURON

The neuron comes in many sizes and shapes and, therefore, is truly polymorphic. The size can range from that of cerebellar granule cells with a diameter of 6–8 μm to pear-shaped Purkinje cells of 60–80 μm. The neurons are characterized by cytoplasmic, basophilic bodies called Nissl substances that are often used as a criterion for identification. The Nissl substance may not be found in axons but may occur along dendrites.

The neuronal cell body has one or more cytoplasmic projections called processes. The processes can be very long (axons) or relatively short (dendrites). The unipolar neuron has an axon only, and the bipolar neuron has an axon and a dendrite, whereas the multipolar has an axon with many dendrites (Fig. 1).

The axons and the dendrites of one neuron are not continuous with another neuron and so they do not make a continuous network. There

5

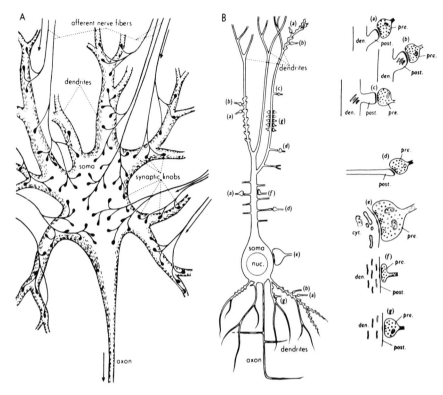

Figure 1. Neuron with synaptic endings. (A) General diagram. (B) Hippocampal pyramidal cell showing diversity of synaptic endings. Magnification of various types of synapses shown on the right. Adapted from Hamlyn (1962) and McGeer *et al.* (1978, p. 8).

is a gap, a characteristic, specialized contact zone of functional inter-neuronal communication, called the synapse. Close to the termination site of an axon there is a large number of spherical microvesicles, the so-called synaptic vesicles; these are not seen in more proximal parts (Fig. 2). The sequence of events occurring at the synapse during signal transmission can be regulated by hormones and drugs and influenced by nutritional stress. These factors will be discussed later.

The neurons are unique in that they exhibit to a great degree the phenomenon of irritability and conductivity. Like all other cells they have fundamental subcellular components such as mitochondria, endoplasmic reticulum, nucleus, golgi, and cytoplasm. The neuroblasts (immature cells) once they are converted to mature neurons in the post-embryonic period do not divide further. It is estimated that up to one-fourth of all neurons (about 30 billion in the human brain) may be lost by the eighth or ninth decade (a daily dropout of about 20,000 cells). This loss may lead to the sensory loss that is common in advancing age, e.g., decreased sensitivity to touch, taste, vision. If the processes of a neuron are destroyed, the cell body may survive but it undergoes certain changes. Some of the cells repair their damage but others will eventually die. Death is more likely if the injury to the axon is closer to the cell body. The neurons of the peripheral system are more likely to be restored than those in the central nervous system (CNS). If the cell dies and the glial cells fill up the tissue a glial scar is formed which, of course, does

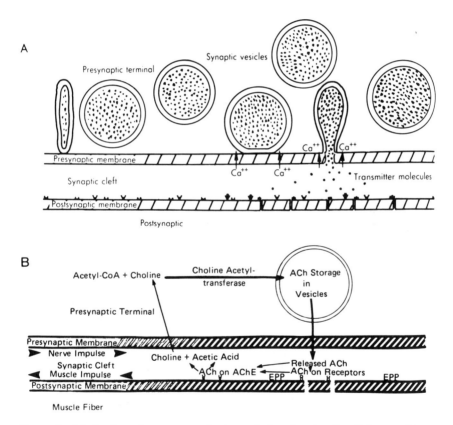

Figure 2. Mechanism of neuromuscular transmission showing the discharge of transmitter molecules (acetylcholine) that then bind with receptor sites on the postsynaptic membrane with the consequent opening up of pores. (A) Movement of CA^{2+} through the presynaptic membrane on depolarization. (B) Schematic representation of the sequence of events in neuromuscular transmission. Reproduced from McGeer *et al.* (1978, p. 92).

not have the impulse-carrying capacity. From a nutritionist's point of view it is very important to realize that in the postembryonic period the neurons generally do not divide and those that die or are damaged are lost forever. There are some exceptions to this generalization. The granular cell layer of the area dentata of the hippocampal region (Altman, 1966), the internal granular layer of cerebellar cortex (Miale and Sidman, 1961), and granular layers of the olfactory bulb (Altman and Das, 1966) are areas where neurons migrate and proliferate after birth, at least in rodents. Regeneration of neurons in the CNS has not been achieved so far but it remains a cherished goal of researchers. Identification of factors that play either a stimulatory or inhibitory role in regeneration is under active research efforts.

Uniqueness of the nerve: All cells have a potential difference across the cell membrane. This has been established by the use of microelectrodes, placed outside and inside the cell, piercing the cell membrane. By passing negative ions into the cell via the electrode, the inside can be made more negative or hyperpolarized, and by passing a current of opposite polarity it can be depolarized. If a depolarizating current is applied, the difference in potential passively follows the imposed potential in most cells, e.g., red blood cells. However, when a nerve cell (usually from an invertebrate like a squid, chosen for its large size) is depolarized from its resting potential of approximately -70 mV to approximately -10 or -15 mV, an explosive change occurs. The transmembrane potential is not only reduced to zero but overshoots zero and may approach $+30$ mV. Thus the inside becomes actually positive with respect to the outside. This process is self-limiting and occurs very rapidly. This is also called the action potential of the nerve cell.

1.2. NEUROGLIA

The word *glia* is derived from "glue" because originally it was thought that these cells were merely supporting structures, holding the neuronal network together. Of course all this has now changed and specific functions have been attributed to the glial cells. Neuroglia is a collective name for all cells that include:

1. Glial cells: (a) astrocytes, (b) oligodendrocytes, (c) microglia (phagocytic cells).
2. Ependyma cells (cells that line brain ventricles).
3. Satellite or capsular cells (supporting cells).

Unlike the neurons the neuroglial cells are nonexcitatory cells; neuroglia can divide throughout life and so deficits can be made up even in later life. The neuroglial cells have a few short processes, at best, and are not involved in signal generation or propagation.

Glial Cells: The glial cells clearly outnumber the neurons (there are seven to eight times more glia than neurons). The oligodendroglia and astroglia have the same ectodermal derivation as neurons. The oligodendroglia have rounded nuclei and are involved in the formation and maintenance of the myelin sheaths of nerve fibers and thus are found to occur in larger numbers in the white matter compared to the gray matter. Nutritional and environmental stresses can affect the myelination process and/or the composition of the myelin. And since any defect in myelin has far-reaching repercussions on brain function, it is essential to know a little more about this vital process.

Myelination: The process of myelination can be described as the wrapping of the axon by the glial cells in the CNS and by the Schwann cells in the peripheral nervous system. The myelin sheath is interrupted by nodes of Ranvier. The process is shown in Fig. 3.

Myelination starts with the initial wraparound by the Schwann cell forming a double membrane. The process continues when the mesaxon

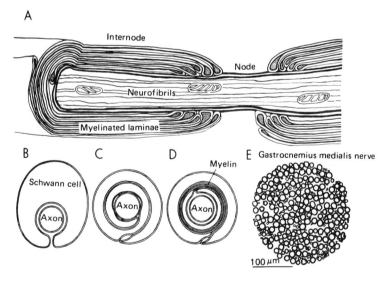

Figure 3. Myelinated nerve fibers. Sequence of myelin formation and Schwann cell wrapping around the axon. Adapted from McGeer, *et al.* (1978, p. 51).

winds around the axon, finally squeezing out the cytoplasm to form compact myelin. In the peripheral nerve myelination is an extension of the double membrane of the Schwann cell (Green, 1954). The process is somewhat similar in the central nervous system, except that the glial cell replaces the Schwann cell (Davison and Peters, 1970). Myelination does not proceed at the same time in all areas of the nervous system and so no general time limit can be given for the myelination of the brain as a whole. It starts in the peripheral system first, followed by the myelination of the spinal cord, and the brain myelinates last. Not all nerve fibers are myelinated. The starting period of myelination could be pre- or postnatal depending on the species. For example, in the guinea pig it is almost over at about birth, whereas in the rat it is a definite postnatal event beginning at about 10–12 days after birth. In humans the maximum rate is observed during the perinatal period and by about 18 months after birth most of the myelin is already laid down. It should be pointed out here that although the peak of the myelination period has been determined in many species, its end is not an abrupt event in any species; in other words, it continues slowly for a considerable period.

Generally, the observation that animals that are born with most of their myelin already deposited are capable of higher complex activity (e.g., horses, cows, sheep) when compared to animals such as kittens and rats, where myelination is a distinct postnatal event; this indicates the importance of proper myelination. From a nutritionist's point of view, the period of peak myelination and timing of the nutritional stress becomes extremely important. Once myelination is complete nutritional aberrations in the later period seem to have only a marginal effect, if any.

The isolation of myelin involves density gradient ultracentrifugation (Norton and Podulso, 1973) and the criteria for purity include electron microscopy (maintenance of a typical five-layered structure with a repeat period of about 120 Å). A marker enzyme, 2′,3′-cyclic nucleotide 3′-phosphohydrolase, has also been used by several workers (Kurihara et al., 1970). At 15 days of age the amount of myelin obtained from one rat brain is only about 4 mg but this increases to as much as 60 mg at 6 months.

Composition of the myelin is characterized by low amounts of water, relatively smaller amounts of protein but a large amount of lipid. The 15–30% protein is comprised of (1) basic protein, (2) proteolipid protein, and (3) a high-molecular-weight protein soluble in acidic chloroform : methanol (Wolfgram protein). The basic protein is unique with respect to its antigenic properties. When injected into animals, it produces an autoimmune disease called experimental allergic encephalitis

(EAE). Animals so affected are sometimes used as models for the study of the demyelinating human disease, multiple sclerosis (Eylar, 1972). The 60–80% lipid is rich in cholesterol, glycolipids (cerebrosides and sulfatides) and plasmalogens (1-alkenyl-2-acyl-*sn*-glycero-3-phosphory-lethanolamine). The molar ratio of cholesterol : phospholipid : galactolipid is 4 : 3 : 2. Other characteristics include the occurrence of hydroxy fatty acids and very-long-chain fatty acids (Pakkala *et al.*, 1966).

The metabolic stability of myelin has been a topic of discussion for several years. In the past, myelin was believed to be inert with almost no metabolic activity. One of the reasons was the early work which showed that myelin cholesterol maintained its radioactive label in the same position even after a prolonged period of incorporation (Davison *et al.*, 1959). Thus, the metabolic activity of one of the compounds that is a major constituent of the brain myelin was found to be very low. In addition myelin did not seem to have any significant enzyme activity other than the cyclic nucleotide phosphohydrolase. All these findings led to the belief that myelin is more or less inert and possibly turns over as a unit. Both of these ideas needed to be modified when Smith (1968) showed that each lipid component of myelin had an individual turnover rate. Three myelin lipids, phosphatidylinositol, phosphatidylserine, and phosphatidylcholine, had half-lives of 5, 4, and 2 months, respectively, whereas cholesterol, sphingomyelin, cerebrosides, and sulfatides turned over in 7–12 months. The brain cholesterol pool is also not homogenous (Kabara, 1973); in some compartments turnover may be fast and in others extremely slow. Myelin turnover rates in general seem to be slower compared to other membrane turnover rates.

The question of turnover of myelin components when there is a lack of enzymes associated with the myelin has not been answered satisfactorily, except to assume that there is a constant exchange of components with other membranes. According to Davison (1972), using 7-dehydrocholesterol as an example, compounds can migrate from myelin, be transformed in the microsomes, and be reincorporated into the myelin. Thus, although the entire process is not clearly understood, metabolic activity must be proceeding even in mature myelin.

2

GROWTH CHARACTERISTICS OF THE BRAIN

2.1. PHYSICAL AND CELLULAR GROWTH

Considering the many observations that have been made to assess brain growth in terms of morphology and biochemistry, one can conclude that the growth of the brain is over long before the rest of the body stops growing. This, of course, does not include the psychological development of the brain, which continues up to and sometimes beyond adulthood.

Table 1 compares some commonly used physical parameters of growth from birth to 3 years.

Along with this physical growth, a complex network of neurons, axons, dendrites, glial cells, and synaptic junctions has already evolved during this growth spurt period; this results in the complex development of physiological and biochemical functions of the brain. Another unique feature is that the neuronal cell population has already reached its maximum number at birth in most species, although in certain small regions of the brain neuronal growth continues beyond birth as mentioned before.

It is important to know that in relation to birth the timing of maximum brain growth varies in different species. The gain in weight of the brain reaches a plateau very early in life and so it is customary to compare the weight gain as a percentage of adult weight. Thus, if one plots the

Table 1. Growth of Human Body and Head Circumference*

Age	Body weight	Total height	Head circumference
At birth	5.7[a]	30	63
At 1 year	16.3	44	83–84
At 3 years	24.0	57	90 or more

[a] All values are given as percentage of adult.

weight of the brain as a percentage of adult weight during the pre- and postnatal growth period the curves are not identical for all species (Dobbing, 1972). This is illustrated in Fig. 4.

One can see that in the guinea pig the peak brain growth occurs in the gestational period, whereas in the rat it is a definite postnatal event. The growth peak in humans seems to coincide approximately with birth. So birth is a significant event for brain growth in humans but not so for either the guinea pig or the rat.

In the case of the rat (the most commonly used laboratory animal model) the first period of cell proliferation and differentiation is complete

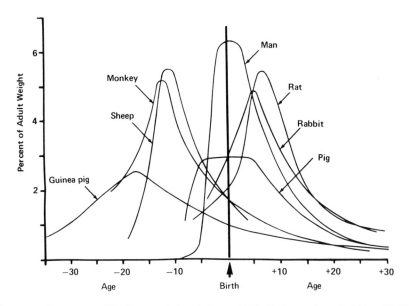

Figure 4. Rate curve of brain growth in relation to birth. Redrawn from Dobbing (1972).

at birth since the number of cells has almost reached adult values. This is the period of mostly neuronal growth. Functionally, nerve impulse transmission does not seem to take place during this period. Next, there is an increase in the size of brain cells with an outgrowth of a network of axons and dendrites. This period corresponds to about the first 10 days after birth. (In humans this would be around birth.) The growth of glial cells and the myelination period averages from 12 to 25 days in the rat, whereas in humans this peak period stretches from birth to about 120 days.

Since most of the brain tissue contains diploid cells and, in the rat, each of these cells contains 6.2 pg of DNA, it is possible to calculate accurately the total number of cells by determining the total DNA of any region in the brain and dividing it by 6.2. In the case of humans this can be done by dividing the total DNA of the region by 6.0 pg. Although this method has been widely used to calculate the total cell number at various stages of brain development, it must be remembered that it does not differentiate between the growth of one cell type and growth of another cell type. Fortunately, the glial cell multiplication peak is a postnatal event, whereas the neuronal peak is prenatal or coincident with birth (in the case of the rat). Once the cell number is determined, other biochemical parameters such as total protein, RNA, and lipid can be expressed as values per cell.

These parameters of brain growth are very useful in interpreting data on the effect of nutritional or other stresses so long as one is aware of the limitations of equating DNA with cell numbers. In the case of an organ that has only one type of cell or in which a particular cell type dominates, the interpretation of data is much more straightforward and meaningful. But in the case of the CNS, many types of specialized cells of different characteristics make up the total mass and this makes the interpretation of data much more difficult. Growth of any organ at early stages can be ascribed solely to cell division and increase in number. Subsequent growth can be due to an increase in size and, therfore, an increase in protein. At maturity one can expect a steady-state characteristic of balanced synthesis and breakdown. At the early stage, DNA data are of great value.

Dobbing (1971) has indicated that there are two separate periods of rapid cell multiplication, one for neurons and the other for glia, giving a bimodal curve (Figure 5). Along with this bimodal curve obtained by DNA determinations Winick (1976) has found that the DNA polymerase activity (responsible for DNA synthesis) also shows a biphasic increase during growth of the brain. To complicate matters further, in various regions of the brain each of these rapid cell multiplication periods occurs

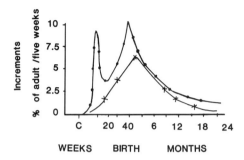

Figure 5. Bimodal curve of DNA, the first one for neuroblast and second for glial multiplication. ●, DNA; ×, cholesterol. Redrawn from Dobbing (1971).

at a different time. Finally, peak myelination does not always coincide with the maximum rate of glial cell division.

One can immediately see the importance of these growth characteristics of the developing brain when one tries to correlate the effects of nutritional stress on the CNS. One can generalize that the timing of nutritional stress would affect the event that is taking place at peak activity at that time. The impact of this phenomenon would be very serious if the cell deficit became permanent with no capacity to correct or ability to restore the balance.

2.2. BIOCHEMICAL PARAMETERS OF BRAIN GROWTH

The rapid changes occurring in the brain have been termed the "growth spurt," "critical period," or "vulnerable period" by various researchers in this field. This period would undoubtedly be the most important for a study of the effects of nutritional and environmental stresses. The following chemical changes that normally occur during this period could be the targets for evaluation of nutritional and environmental impacts.

2.2.1. Water Loss

During the earlier period when cell division is dominant the brain tissue contains about 90% water and 10% solids. During the subsequent entire period of rapid growth of the brain there is a continuous decrease in brain water. Rapid increase in the number of cells and their processes in a limited space is one of the causes of this decrease. The other cause is the process of myelination. The cytoplasmic water in the glial cell which wraps around the axon gets squeezed out as the wrap gets tighter. The loss of water begins at about 7 days and continues over a very long period. However, the rapid rate of decrease is over by about 60–70 days. The adult brain contains about 75% water and 25% solids out of which only one-tenth is made up of water-soluble components. During the growth and development of the CNS there is a redistribution of intra- and extracellular water. In the case of an adult rat cerebral cortex, the extracellular space is predicted to be about 14.5% (Himwich, 1973; McIllwain and Bachelard, 1971; Tower, 1965).

2.2.2. DNA, Protein, and Lipids

Whereas DNA in the brain is indicative of cell number, lipids (especially some specific components) are generally considered an index of myelination, and proteins make up the major portion of the remaining solids (glycogen–carbohydrate content of the brain is very low). Thus these three constituents have received much attention during the period of active growth and development of the brain. Figure 6 shows the increments in these components as a percentage of adult values.

Figure 7 shows that in less than 3 weeks' time, the maximum accumulation of the three components, DNA, protein, and lipid, has occurred in the developing rat brain. The curve for protein does not level off like the DNA curve because the protein content of the cell increases along with the increasing size of the cell, indicating the stage of hypertrophy, even though the stage of hyperplasia is already over. The lipid content of the brain increases approximately threefold from birth to 4 weeks. Since myelin is composed of 80% lipid, progress in myelination culminates in a rapid increase in brain lipid per gram of wet brain tissue. Brain lipids are much more complex than the lipids of liver or adipose

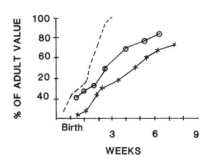

Figure 6. Increase in DNA, protein and lipid in the rat brain during development. Values are expressed as a percentage of maximum adult values. – – –, DNA; o——o, protein; ×——×, lipid. Redrawn from Benjamins and McKhann (1976).

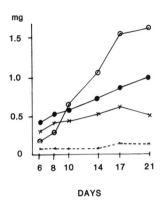

Figure 7. Total DNA content in various regions of the rat brain. ○——○, Cerebellum; ●——●, cerebrum; ×——×, brain stem; ●– – –●, hippocampus. Redrawn from Winick (1976, p. 41).

tissue. One would expect variations in many individual components of this complex mixture. Such variations will be considered in a separate section.

The DNA curve shown in Fig. 6 is representative of the total brain DNA; but, as pointed out earlier, all the regions of the brain do not develop simultaneously and, therefore, each region would have a DNA curve of its own. This is depicted in Fig. 7. The growth of the rat brain stem is already over by 17 days but the increase continues in the cerebrum. The curve for the cerebellum shows a very steep rise up to 17 days and then starts to plateau. The curve for the hippocampus is unique in that the growth is negligible up to 14 days but shows a significant increase in the next 3 days and flattens out again. The precise increase between the 14th and 17th day, however, is not due to an increase in the rate of cell division but rather due to a migration of neurons into the hippocampus from under the lateral ventricle (Altman, 1966; Altman and Das, 1966).

There are two ways to look at this phenomenon in relation to the effects of nutritional stresses. One way is to determine the rate of cellular growth and the maximum slope of the DNA increase curve, and theorize that malnutrition during this time period will have a most damaging effect. The other way is to look at a specific phenomenon such as myelination, which may not coincide with the maximum rate of cell differentiation, determine the period at which it is occurring at a maximum rate, and conclude that this time period is most vulnerable to nutritional and environmental insults. However, the author feels that to evaluate the effects of malnutrition on the developing brain, both of these points of view should be considered together.

2.2.3. Biochemical Changes during Development

Measurements of changes during development need not be confined to structural constituents but should include enzyme activity, hormonal activity, and other metabolic activities such as utilization of oxygen. Among these various parameters the metabolism of neurotransmitters that regulate signal transmission may be considered of prime importance because they relate to functional activity of the brain. During development of the CNS and as neuronal processes, synaptic junctions, and myelination evolve, the functional components also increase rapidly. Figure 8 shows the levels of norepinephrine (NA) and serotonin (5HT) and the levels of monoamine oxidase (MAO), the enzyme intimately connected with the metabolism of these neurotransmitters. The levels of neurotransmitter compounds as well as the activity of MAO increase rapidly up to about 30 days and then start leveling off. On the other hand dopamine, another well-known neurotransmitter, showed very little change per gram of wet brain (Hyyppä, 1971).

Another example of enzyme activity changing during brain development is provided by the enzymes of the hexose monophosphate (HMP) shunt that metabolize glucose and produce NADPH necessary for lipid synthesis. During the first 3 weeks of life, these enzymes play a major part in the rat, but, after maturation, the Embden–Meyerhof (EM) pathway takes over the major task of glucose utilization (Winick, 1970).

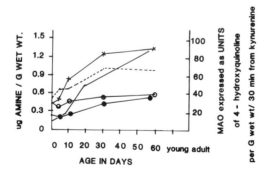

Figure 8. A comparison of the levels of NA and 5HT in hypothalamus and cerebral cortex of rats at different ages as related to the levels of MAO activity. ●——●, NA cortex; ○——○, 5HT cortex; ——, MAO activity; ×——×, NA hypothalamus; – – –, 5HT hypothalamus. Redrawn from Rockstein (1973, p. 81).

Incorporation of radioactive acetate into brain total lipids (Dhopeshwarkar *et al.*, 1969), an index of lipogenesis in the brain, is also very high during the active growth period and levels off as maturation is attained. The fatty acid synthetase activity (Volpe and Kishimoto, 1972) was found to be very high in fetal brain and when expressed per brain there was a steady decrease in activity during the suckling period. However, the authors point out that the fall in synthetase activity was considerably less compared to the weight increase of the brain during the same period and thus the total activity of fatty acid synthetase actually rose during the suckling period. Since the fatty acid synthetase activity was highest in fetal tissues, obviously it has no relation to the peak period of myelination in the rat. The persistence of enzyme activity throughout the growth period indicates the importance of lipid buildup in the brain.

Another enzyme that is influenced by the age of the rat is the cholesterol esterifying enzyme (Eto and Suzuki, 1972a; Jagannatha and Sastry, 1981). Cholesterol esters are found in relatively larger amounts before myelination than after but in the adult brain only a negligible amount can be detected (Alling and Svennerholm, 1969; Eto and Suzuki, 1972a). This led to a search for the esterifying enzyme. The enzyme characterized by Eto and Suzuki (1972b) had a pH optimum at 5.2, whereas the one recently reported by Jagannatha and Sastry (1981) showed a more physiological pH optimum at 7.4 and also a dependence on ATP and coenzyme A (CoA). These enzymes have also been associated with myelin. The pH 5.2 enzyme increased up to the 6th postnatal day and then remained nearly constant throughout adult life; however, the pH 7.4 enzyme increased markedly just prior to the period of active myelination reaching a peak at the 15th day and subsequent decrease as the rat aged. The pH 5.2 enzyme explains the occurrence of cholesterol esters in the brain during the early period before myelination but the implication of the other enzyme (peak at myelination period) remains obscure because cholesterol esters are normally not found in large amounts in the brain at this stage.

The enzyme glutamic acid decarboxylase (GAD), which is responsible for conversion of glutamic acid to γ-aminobutyric acid (GABA), increases tenfold in the rat cortex from birth to maturity (Sims and Pitts, 1970). During this period, therefore, the twofold increase in GABA is understandable, but, contrary to expectation, the glutamic acid level increases rather than diminishes (Agrawal *et al.*, 1966)! When dealing with enzyme activities in a whole-brain preparation only a very gross picture is seen. Since the enzyme activities would not be expected to be uniform in all areas of the brain, the anomaly cited above may not be difficult to explain. Both the enzyme activity distribution and variation in activity could very well result in an unequal distribution of product in the various areas of the brain. This was very well illustrated by Kandera et al. (1968) who found that there was six times as much glycine per gram of wet weight of tissues of pons and medulla of adult rat as there is in the cerebral or cerebellar hemispheres. On the other hand, a reverse relationship was found in the case of taurine and glutamic acid. The concentration of brain glutamic acid and that of the branched-chain amino acids phenylalanine and proline shows exactly opposite trends. The former increases but the latter decreases (Agrawal et al., 1966; Bayer and McMurray, 1967). The dramatic change in the free amino acid pool is not confined to the active growth in the suckling period as shown by Lajtha and Toth (1973). These authors report a decrease of about 50% in lysine and 20% in alanine (in mice) in the first 24 h of life. Taurine levels are highest at birth, slowly decreasing to about 40% during the growth period.

The increase in the free glutamic acid content of the brain during development has been repeatedly observed by many workers. But the mechanism of this increase is not entirely clear. It could be due to an increased rate of de novo synthesis (Gaitonde and Richter, 1966). Another possible explanation revolves around increased uptake during the developing period. Such an increased uptake was found only in neonates (Himwich et al., 1957).

2.3. MORPHOLOGICAL CHANGES

As mentioned earlier, after the initial stage of hyperplasia, when the volume of the cell increases, the stage of hypertrophy of the cell is established. One can expect that, along with size, the complexities of the neuronal processes also increase (Eayrs and Goodhead, 1959; Aghajanian and Bloom, 1967). These authors have studied both the molecular layer of sensorimotor cortex for axonal density and the synapse (using electron microscopy to distinguish cell volume from increased dendritic growth) and have shown that both these parameters increase during rapid brain development (Fig. 9).

In conclusion, one can say that the developing brain is in many ways different from the mature one. During development, cell number, composition, and complexity of morphological structure are constantly changing, whereas in the mature brain these parameters are for the most part stable. The period of development is relatively short and full of rapid changes. The initial phase is characterized by neuronal prolifera-

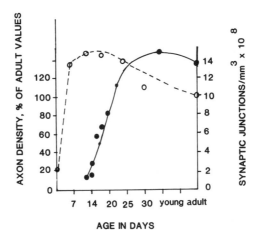

Figure 9. A comparison of the increase in axon density with the increasing number of synaptic junctions observed in layer 1 of the cerebral cortex. o– – –o, Axon density sensorimotor cortex; ●——●, synapses, parietal cortex. Redrawn from Ford (1973, p. 83).

tion and the later stage signifies glial proliferation and myelination. Both of these stages can be affected by nutritional and environmental stresses and there may not be another chance for remedial correction. The adult brain is far more protected and effects are much less severe if similar stresses are imposed.

So far we have restricted this discussion to biochemical parameters and have not included psychological development. This subject is beyond the scope of this book and the reader is referred to other monographs that include this aspect (Dodge *et al.*, 1975).

3

ENERGY SOURCE

3.1. GLUCOSE

The source of energy for the various growth characteristics and parameters described in the previous chapter originates from the available supply of nutrients. Among these nutrients, except for short periods of time, glucose and oxygen seem to be the most important. Both of these nutrients are considered obligatory requirements and not just preferred sources of energy. Without an adequate supply of these two substances, the brain cell can neither develop nor survive. Long before current ideas of metabolism were known, this conclusion was reached on the basis of the simple fact that even short disruptions of blood supply to the brain were either fatal or produced long-lasting damage. Even in the presence of adequate blood circulation, severe hypoglycemia produced similar effects. When ideas about basal metabolic rate and respiratory quotient were developed, it was soon found that the respiratory quotient (RQ) of the brain (the ratio of amount of CO_2 produced to amount of O_2 consumed) was close to one. It was well known that utilization of lipids gave a RQ of 0.7; this confirmed that carbohydrates were the major source of energy fuel for the brain. The obligatory nature of the requirement of glucose was further provided by the fact that hypoglycemia produced by the administration of insulin altered most parameters of brain function and these could not be restored by any compound other than glucose.

During the early period of development, the energy is utilized for synthetic processes as well as normal functions associated with the brain. However, on reaching maturity, the synthetic process decreases in magnitude and so the required energy is mostly channeled toward "brain functions." For example, a 10-day-old rat uses glucose in the ratio of 5 moles to 1 mole of oxygen, whereas a 30-day-old rat uses only 3.9 moles of glucose per mole of O_2 utilized. In human studies, it was found that in childhood the cerebral blood flow is 106 ml/100 g/min, in young adulthood it is 54 ml/100 g/min, and the requirement of adult human brain is calculated to be about 4.6 g of glucose per hour (Ganong, 1971).

Several substances, amino acids and intermediates of carbohydrate metabolism, have been tested in an effort to determine if they can substitute for glucose. However, most of them are effective only because they raise blood glucose values. For example, epinephrine and maltose are useful substitutes only because they help to increase blood glucose concentrations. Several intermediates, such as lactate, pyruvate, fumarate, acetate, and glyceraldehyde, can be metabolized by the brain but are ineffective due to either low concentration in the circulating blood or the inability to easily pass the blood–brain barrier. Mannose, which is not a major constituent of diet or of blood plasma, seems to be the only effective substitute which is directly metabolized to produce energy. β-Hydroxybutyrate and fatty acids are considered alternate sources of energy and as such will be considered in detail as they need special understanding of the changing composition of the diet during immediate postnatal development. Finally, it should be remembered that just because the brain contains all the necessary enzymes for the utilization of certain metabolic intermediates or dietary components, it does not mean that these compounds can replace glucose requirements of the brain *in vivo*. Other factors, such as concentration in the blood, uptake by the brain, and concentration of enzymes for metabolic reactions, have to be taken into account. The levels of oxidative enzymes increase from birth to completion of the brain growth spurt in the rat but thereafter there is very little change. The high enzymatic activity is needed for the dual purpose of utilizing the compound for energy and providing high synthetic activity in the developing brain.

The supply of blood (O_2 and glucose that is carried with it) is sometimes decreased in the normal course of pregnancy to one of the fetuses in animals delivering a litter of pups and this results in a "runt" in the newborn progeny. These undernourished newborns usually do not survive to reach maturity. To study this phenomenon, researchers have conducted experiments in which the uterine artery to one of the horns of the uterus is ligated thereby reducing the blood supply only on one side. At birth all the pups from the side of reduced blood supply show decreased weights. Ligation during the period of rapid cellular growth will reduce the cell number, whereas the same treatment later in the period will affect the cell size. It is interesting to note that when ligation type experiments were conducted, all fetal organs were not affected to the same extent; liver seemed to be affected the most but the brain was spared to a greater extent than any other organ (Winick et al., 1972a). Such a phenomenon is sometimes observed in small-for-date human infants. The increase in blood flow in different regions of the brain follows the growth and development of that region.

There are two sources of glucose available to the brain. One is the blood glucose and the other is brain glycogen. The former, maintained within a close concentration range, is the larger of the two sources. The concentration of glycogen in the brain is not very high. For example, rat liver contains 3.3–4.5 g of glycogen per 100 g, muscle contains 0.3–0.45 g/100 g (the total mass of muscle is significantly greater than that of liver), and brain contains only 0.092 g/100 g. It should be pointed out here that within 3 min after death, up to 50% of brain glycogen is hydrolyzed to glucose and so special precautions and techniques are needed for the determination of glycogen in tissues (McIllwain and Bachelard, 1971). The brain glycogen is mainly reserved for emergencies and not for normal day-to-day needs. It is isolated from other systemic demands. Epinephrine is known to stimulate glycogenolysis, but, in normal circumstances, its uptake by the brain from circulating blood is very much restricted by the blood–brain barrier system.

The energy from glucose itself can be derived by two distinct metabolic pathways. The first is glycolysis via the EM pathway followed by oxidation of pyruvate by the tricarboxylic acid (TCA) cycle and the second is direct oxidation of glucose by the HMP shunt. Mention has already been made of the fact that approximately half of glucose metabolism occurs via the HMP pathway during active development and growth of the rat brain (up to about 30 days). The biggest advantage of using the HMP pathway—direct oxidation of glucose—is that it generates the NADPH required to sustain rapid lipid synthesis for the myelination process and generates ribose molecules for nucleic acid metabolism. Generally, in adults 90% of the glucose undergoes glycolysis via the EM pathway with the pyruvate being further metabolized by the TCA cycle to produce ultimately CO_2 and H_2O and the energy that is derived is trapped in the form of the generated ATP. A small amount of lactate is also produced from the pyruvate. During periods of oxygen deprivation, the only energy available to the brain is from energy-rich phosphate bonds of ATP and phosphocreatine but this source is short-lived because all such sources are depleted within minutes and thus are unable to sustain long-term physiological functions. Relative to the adult brain, the newborn brain is able to withstand longer periods of hypoxia but this is not without some damaging effect. For example, it has been shown that the rate of cell division is affected by hypoxia. Hypoxia can occur during gestational periods as well as early postnatal periods due to certain complications. Before birth, glycolysis is a predominant process but during the postnatal development respiration assumes the dominant role.

Glucose entering the brain not only serves as a source of energy but it also provides carbon for synthesis of other important compounds. For example, in the rat, between the ages of 10 and 30 days, there is a great rise in the conversion of glucose carbon to amino acids of the cerebrum (Patel and Balazs, 1975). When glucose is oxidized by the EM pathway and the resulting pyruvate enters the TCA cycle, α-ketoglutarate and oxaloacetate are produced. By converting these two compounds to glutamate and aspartate, respectively, glucose carbon can be trapped to form amino acids. In fact, the concentration of both glutamate and aspartate increases during periods of rapid growth and development. This parameter has been adapted as a measure of the effects of malnutrition on the CNS (Balazs, 1973). Balazs found that conversion of glucose carbon into amino acids was severely decreased in malnutrition. However, this effect was reversible by adequate diet, i.e., the flux of glucose carbon into amino acids was restored but whether this reversal corrected the "damage" already caused remains unknown.

3.1.1. Glucose versus Fructose

One of the misconceptions prevalent in certain populations is that the nutritional value of fructose is somewhat better than that of glucose. Fruit sugar or honey is wrongly preferred over sucrose. This misconception may have its origin in the fact that diabetics are able to clear blood fructose but not glucose. Individuals seeking weight reduction also prefer fructose over glucose. But this may be explained only on the basis of relative sweetness. Taking sucrose as a standard (100) the relative sweetness of glucose is 74 whereas that of fructose is 173! Therefore, less fructose is needed to attain the same amount of sweetness. This decreased amount may, under ideal conditions, lead to weight loss. Biochemically there can be no difference between glucose and fructose because glucose is converted to fructose phosphate in the glycolytic pathway. It appears that brain and also muscle can utilize significant quantities of fructose only after the conversion to glucose. In the CNS the uptake of fructose from the circulating blood is considerably less than that of glucose. For example, if the uptake of water is given a value of 100, D-glucose has an index of 33, whereas D-fructose has a value of only 1.75 (Oldendorf, 1971). Similarly, in the intestine the absorption of glucose far exceeds that of fructose (due to absence of active transport for the latter), and, therefore, the blood fructose does not rise as rapidly as glucose. Thus, there is no reason to believe that honey (a rich source of fruit sugar—fructose) is in any way superior to glucose or sucrose.

3.2. ALTERNATE SOURCE OF ENERGY (KETONE BODIES)

There is an abrupt change in nutrient availability for the fetus as compared to the newborn rat; the fetus derives its nutrient supply from the maternal circulation, which is relatively rich in carbohydrates (blood glucose). However, soon after birth, the pup starts suckling the initial milk (colostrum) supply that is extremely rich in fat and relatively poor in carbohydrates. [Rat milk contains 14.8% fat which is approximately three times higher than that of human milk (Cox and Mueller, 1937).] Thus, the newborn rat life system is faced with an abrupt change in dietary constituents. The body then tries to quickly adapt to use this dietary fat for all its requirements. The maternal blood during gestation usually contains a higher amount of fat, mostly triglycerides; however, it is not known whether an increased amount of lipids (mostly free fatty acids) are transported into the fetus. From these observations, it seems that the rat fetal organs, including the brain, either have an enzymatic system for utilization of lipids for survival or develop such a system very rapidly.

The utilization of lipids starts with liberation of fatty acids from complex lipids followed by β-oxidation of fatty acids to produce acetyl-CoA. Two units of acetyl-CoA can condense to form acetoacetyl-CoA and in the liver a thiolase converts these to free acetoacetate. This keto acid is reduced to β-hydroxybutyric acid or is decarboxylated to form acetone. These normal breakdown products, collectively termed ketone bodies, are released into the circulation. Acetone is excreted in the urine and exhaled in the breath. Tissues other than liver, e.g., muscle, transfer the CoA group from succinyl-CoA to acetoacetate to form the CoA ester acetoacetyl-CoA. This activated form is oxidized to CO_2 and H_2O via the TCA cycle to generate adenosine triphosphate (ATP) and energy. The normal blood ketone level in man is about 1 mg/100 ml and almost all of it is rapidly metabolized, thus preventing accumulation in the liver and, in turn, in the circulating blood.

The initial argument that CNS function can be maintained only in the presence of glucose was proposed after inducing hypoglycemia using an insulin injection. In these circumstances, lack of glucose quickly caused CNS dysfunction, sometimes followed by coma. As soon as glucose was made available, the effects were reversed, and normal CNS function was restored. However, injection of insulin, because of its antilipolitic effect, also caused a decrease in utilization of plasma free fatty acids (FFA) and ketone bodies (Williamson and Buckley, 1973). Because this fact was overlooked, even though earlier work by Quastel (1939) had suggested that brain tissue can oxidize ketone bodies, the suggestion was ignored. Owen et al., (1967) have reported that in patients undergoing prolonged starvation as therapy for reducing obesity, there was a considerable uptake of both acetoacetate and β-hydroxybutyrate by the brain, as judged by arteriovenous differences. It was soon concluded that in the adult this uptake of ketone bodies was a reflection of their higher concentration in the circulating blood rather than a change in enzyme concentration (Williamson et al., 1971). However, when neonatal or very young rats are considered, an entirely new concept becomes apparent.

There is reason to believe that the concentration of plasma free fatty acids is higher in young suckling rats than in adults. The main reason for this is that during the transition from intrauterine to extrauterine life the rat faces an abrupt change in composition of nutrient supply. The high fat in the rat milk necessitates elevation of levels of enzymes related to ketogenesis (Lockwood and Bailey, 1971). Itoh and Quastel (1970) have studied the oxidation of glucose and acetoacetate in brain slices from infant and adult rats. First, they noted that the rate of $^{14}CO_2$ formed from $[1 - {}^{14}C]$glucose far exceeds that from $[6 - {}^{14}C]$glucose in respiring brain slices from infant rats as compared to adult rats. This suggests that the HMP shunt oxidative pathway is more active in the immature rat brain. Second, they found that acetoacetate or β-hydroxybutyrate is a more potent source of acetyl-CoA than glucose, in the infant brain slices. This was attributed to a decreased rate of operation of the TCA cycle in such tissue. Finally, a decreased rate in the utilization of glucose carbon for synthesis of amino acids, mostly glutamate and aspartate, was noted in the infant rat brain. Reactions involved in ketone body utilization are as follows:

1. D-3-Hydroxybutyrate + NAD $\rightleftarrows$ acetoacetate + NADH + H$^+$ catalyzed by the enzyme hydroxybutyrate dehydrogenase (EC 1.1.1.30).

2. Acetoacetate + succinyl-CoA ⇄ acetoacetyl-CoA + succinate catalyzed by 3-oxoacid CoA transferase (acetoacetyl-CoA transferase) (EC
2.8.3.5).
3. Acetoacetyl-CoA + CoA ⇄ 2-acetyl-CoA catalyzed by acetoacetyl-
CoA thiolase (EC 2.3.1.9).

Klee and Sokoloff (1967) and Dahlquist et al. (1972) have shown that the
enzyme β-hydroxybutyrate dehydrogenase, which catalyzes the first reaction, has an activity three to four times higher at age 15 days than at
180 days. They also found a steady increase in its activity up to 30 days
and then a gradual leveling off as the rat aged. This activity was confined
to the brain mitochondria, but the decrease in activity as the animal aged
was not due to a change in permeability of the mitochondrial membrane.
Lockwood and Bailey (1971) have shown that the activity of 3-oxoacid
CoA transferase increases steadily after birth and peaks at 30 days.
Dierks-Ventling and Cone (1971) reported that the third enzyme, acetoacetyl-CoA thiolase, is quite active in the fetal brain and also during
the first 3 weeks of postnatal life. Thus, the in vitro studies have shown
that all the necessary enzymes needed for ketone body utilization have
higher activities during the suckling period.

These observations, when considered along with the higher levels
of ketone bodies in rat blood during development (Page et al., 1971;
Dalquist et al., 1972), suggest that the neonatal brain is specially equipped
to utilize this readily available source of energy (ketone bodies). The
liver, where these ketone bodies are formed, is one tissue that cannot
use them metabolically, mainly because of the low level of the activating
enzyme 3-oxoacid CoA transferase. The source of the ketone bodies in
the in vivo situation are plasma free fatty acids, which are elevated during
the suckling period (Page et al., 1971). However, acetoacetate can arise
by two possible pathways. First, catalyzed by a thiolase enzyme, two
acetyl-CoAs can condense to form acetoacetyl-CoA + CoA to be followed by deacylation to give acetoacetate. In the second pathway, acetoacetyl-CoA + H_2O + acetyl-CoA, catalyzed by hydroxymethylglu-
taryl-CoA (HMG-CoA) lyase, can form HMG-CoA and CoA to give free
acetoacetate and acetyl-CoA. This later reaction was found to be the
major pathway in the liver (Williamson et al., 1968).

Further confirmation that ketone bodies are a source of energy to
the developing brain came from in vivo studies using a method of measuring arteriovenous differences. Hawkins et al. (1971) have conclusively
shown that ketone bodies are taken up by the brain of the suckling rat.
Continuing the discussion, Williamson and Buckley (1973) point out that

the enzyme levels for ketone body utilization are the highest in the heart and kidney of adult animals while being lowest at birth. On the other hand, the young rat brain has higher enzyme activity than adult brain. This together with the fact that brain weight is higher than heart or lungs in neonatal rats makes it a major tissue to utilize ketone bodies. Edmond (1974) has pointed out that liver is committed to the production of ketone bodies but cannot use them and, therefore, passes them on to the CNS which cannot make the ketone bodies by itself.

Hawkins and Biebuyck (1979) point out a major difference between utilization of glucose and ketone bodies as a source of energy. Whereas glucose seems to be metabolized freely by all cerebral areas, the utilization of ketone bodies is governed by transport across the blood–brain barrier system. This may explain the failure of ketone bodies (even when circulating concentrations are high) to maintain electroencephalographic activity during phases of induced hypoglycemia (glucose concentrations lower than 1 to 2 mM) (Zivin and Snarr, 1972). Since the ketone body concentration in the brain tissue itself is very low, the source has to be the circulating blood. It is well known that the ketone bodies are not direct products of fatty acid degradation but are synthesized and regulated by separate pathways. During fasting, in diabetes, during pregnancy, during the perinatal period, or during consumption of a high fat diet, the concentration of ketone bodies in the blood is increased. However, this increased amount of ketone bodies in the blood cannot readily pass the blood–brain barrier. Therefore, in areas of the brain where the uptake is relatively less restricted, one can expect a higher utilization. Hawkins and Biebuyck confirmed this (1979) by injecting labeled hydroxybutyrate and examining uptake by autoradiographic techniques. The transport of ketone bodies from the blood into the brain is governed by a carrier-mediated transport mechanism; several acids, lactate, pyruvate, acetoacetate, and β-hydroxybutyrate, compete for this transport mechanism. During fasting, lactate and pyruvate (carbohydrate metabolites) decrease, and so the higher amount of circulating ketone bodies can easily pass into the brain. This may be an explanation of how the brain is able to use ketone bodies for energy purposes during fasting, and Owen et al. (1967) have found that 60% of the energy requirement comes from utilization of ketone bodies during prolonged fasting.

In summary, one can say that ketone bodies (lipids) are undoubtedly an alternate source of energy for the neonatal brain, and under extremely stressful conditions even adult brain can utilize this source. If it were not possible to use this source and the brain was solely dependent on glucose, then during prolonged starvation all of the body protein would be exhausted since the carbohydrate stores in the body are small and the needed glucose would have to come from protein via gluconeo-genesis. It is hypothesized that certain areas of the brain are impermeable to ketone bodies and, therefore, there is an obligatory need for glucose for the brain.

In human studies, the newborns as well as infants have been shown to have a higher level of ketone bodies in their circulating blood when compared to levels in normal adults (Persson and Tunnel 1971). Using autopsy material, Tilden and Cornblath (1972) have shown that enzyme levels necessary for ketone body utilization are present in adequate amounts in human fetal brain at 32 weeks of gestation as well as in infants and have reported a case of succinyl-CoA : 3-ketoacyl-CoA trans-ferase deficiency. In spite of this supporting evidence, β-hydroxybutyr-ate is incapable of maintaining or restoring brain function following hypoglycemia, either spontaneous or artifically induced. In addition to previously discussed causes, glucose may be necessary to provide ad-equate levels of succinate, via the TCA cycle, which in turn is needed for the acetoacetyl-CoA transferase reaction that is a vital link in the pathway for ketone body utilization. Brain damage in human infants due to untreated hypoglycemia or in small-for-date babies has been observed (Lubchenko and Bard, 1971). However, if the cause of hypo-glycemia lies in insulin-related problems then it involves not only un-derutilization of glucose but also underproduction of ketone bodies, as explained earlier. These two factors must be separated before final con-clusions can be drawn. Using a rat model, Chase *et al.* (1973) have reported the effects of hypoglycemia induced by insulin administration once daily for 18 days after birth. These effects include reduced brain weight, decreased cellularity, reduced protein in the whole brain, and also a lower level of myelin lipids. However, this method of inducing hypoglycemia leaves two possible explanations as mentioned above. The two factors, hypoglycemia and decreased lipolysis, must be separated before any conclusions can be drawn.

BLOOD–BRAIN BARRIER SYSTEM

4.1. EARLY STUDIES

All substances introduced into the bloodstream, either by normal absorption of dietary compounds and the metabolic products produced from such compounds in the body itself or compounds administered for therapeutic purposes, do not enter the brain cell at the same rate or to the same extent. The magnitude of permeability from the blood into the brain is not the same in all areas of the brain and, in certain cases, depends upon the age and the species. This selective uptake of compounds will thus have an important bearing on the nutritional effects on the developing brain.

The phenomenon was first described by Ehrlich (1882) 100 years ago. He found that if dyes were injected intravenously, all the organs were deeply stained but the brain remained unstained. Goldmann's experiment (1913) further proved that injected trypan blue did not stain the brain tissue but the choroid plexuses and meninges were stained. If the dye was injected into the cerebrospinal fluid, the brain tissue was intensely stained because in this method the blood–brain barrier was bypassed. From this point on, the concept of the blood–brain barrier between brain capillaries and the brain cells has become an accepted concept but the scientific explanation for this concept is still a focus of controversy. The strongest and most vocal argument against the trypan blue studies has centered around the fact that immediately after injection such dyes quickly form stable complexes with blood proteins, and what one observes is the inability of the proteins to pass out of the capillaries. However, as Davson (1967) points out, the brain capillaries must be unique because protein-bound stain did come out of the capillaries in all other organs. Factors such as fenestrations, tight junctions between capillary endothelial cells, lack of pinocytosis, and covering of the brain capillaries by astrocyte end feet have been proposed and discussed in relation to the blood–brain barrier by Davson (1967) and the basic concept challenged by Dobbing (1961).

The complexities of this phenomenon make it almost impossible to guess which compound will readily pass into the brain or the criteria of what restricts the passage. A simple explanation like molecular weight, solubility, or charge does not seem to offer a correlation between restricted or unrestricted permeability. Thus with some exceptions, it is almost always necessary to determine experimentally the actual uptake of any given compound by the brain following its introduction into the blood.

There are three major experimental techniques to determine brain uptake. One is to introduce the substance, preferably radioactive, into the carotid artery and collect blood from the superior sagittal sinus (blood leaving the brain directly). In this method the animal need not be sacrificed at the end of the experiment. A second method is to establish a concentration gradient in the circulating blood, and after several intervals sacrifice the animal, remove the brain, and chemically determine the concentration in the brain homogenate. A third method recently introduced by Oldendorf (1971) uses a different approach. A ^{14}C-labeled test compound along with $^{3}H_{2}O$ is injected into the carotid and 15 sec later the animal is decapitated. The amount of test substance remaining in the brain after this short interval (enough for a single pass) is expressed as a percentage of the $^{3}H_{2}O$ in the brain. Since only the ratio of $^{3}H/^{14}C$ is determined, the method is free from complicated chemical manipulations such as extraction and isolation. Since $^{3}H_{2}O$ would be taken up completely, it is given an arbitrary number 100. Using this technique Oldendorf has reported the uptake of 28 amino acids, 13 amines, 7 hexoses, and 5 relatively diffusible substances. He found that essential amino acid uptake was higher than nonessential amino acids. Glucose uptake was stereospecific (D-glucose index was 33 compared to 1.6 for L-glucose). The index for glutamic acid was 3.2 but the index for glutamine was 7.6. Fructose had an index of 1.75. From these figures one can see that a small change in molecular structure or configuration made a remarkable change in the brain uptake. This is precisely the reason why predictions about uptake can only be guesses at best.

Years ago, Davson (1967) had shown that lipid solubility greatly facilitated uptake of substances into the brain, e.g., ethyl urea was taken up in higher amounts than urea. A similar observation regarding anti-pyrene, a lipid-soluble substance, and N-acetylaminopyrene, a much less lipid-soluble compound, was made by Meyer et al. (1959). However, the penetration of cholesterol from the blood into the brain, in spite of lipid solubility, was reported to be very low (Bloch et al., 1943). Bilirubin is a lipid-soluble bile pigment that binds very strongly with plasma proteins. Thus, normally no bilirubin leaves the brain capillaries. In contrast, FFA bind strongly with plasma albumin but this seems to be the preferred form of FFA transport (Dhopeshwarkar et al., 1972).

4.2.   TRANSPORT OF LIPIDS FROM BLOOD INTO THE BRAIN

The discovery of the inability of cholesterol to pass out of the brain capillaries into the brain in spite of lipid solubility refocused attention on lipid-soluble compounds. In our own laboratory (Dhopeshwarkar and Mead, 1973), we were interested in determining whether fatty acids, either orally ingested or injected into the bloodstream, would be taken up by immature and mature animals. Most compounds, but not all, are taken up by the brain of younger animals, and therefore we chose to use full-grown adult animals in our study. The rationale for studying the uptake of fatty acids was as follows: saturated and monounsaturated fatty acids, such as palmitic, stearic, and oleic acids, can be synthesized by the brain from available precursors. Glucose carbon can be a source of brain fatty acids (Dhopeshwarkar and Subramanian, 1977); however, essential fatty acids, linoleic ($18 : 2\omega6$) and linolenic acids ($18 : 3\omega3$), cannot be synthesized by the mammalian system. Thus, these two acids must be provided by the diet just like essential amino acids, vitamins, and minerals. Most mammalian systems will then synthesize longer-chain polyunsaturated fatty acids (PUFA) from these two precursors to build up families of fatty acids called $\omega6$ and $\omega3$ unsaturated fatty acids. The essentiality of $\omega6$ fatty acids has been well established. They are required to maintain the integrity of the cell membranes and also to serve as precursors for the synthesis of biologically important com-pounds, the prostaglandins (Mead and Fulco, 1976). The essentiality of $\omega3$ fatty acids is still under investigation and remains unknown; but it is well known that the brain and retina contain relatively large amounts of ω3-PUFA both in young and adult animals including humans.

If one injects a ^{14}C-labeled fatty acid into the bloodstream and examines the brain and finds the radioactive fatty acid in the brain, one is inclined to conclude that the fatty acid was taken up by the brain. But this kind of conclusion is deceptive and oversimplified. For example, if one starts with $[1-^{14}C]$palmitic acid, upon injection into the bloodstream it may be oxidized to produce labeled acetyl-CoA which, in turn, can enter the brain and be utilized for resynthesis of palmitate. Thus, although radioactive palmitate was found in the brain, there was no conclusive proof of direct uptake of the starting material. This argument holds true for any fatty acid that can be synthesized in the brain *in situ*. In the case of essential fatty acids, linoleic and linolenic acids, the simple experiment as stated above could be conclusive proof since these essential fatty acids cannot be synthesized by the brain or any other mammalian system. A technical difficulty is that both linoleic and linolenic acids are found in extremely small amounts in the brain and therefore are difficult to isolate and purify. However, their products, arachidonic (20 : 4ω6) and docosahexenoic acids (22 : 6ω3), are found in rather large amounts and can be easily isolated.

Taking advantage of label distribution studies, made possible by a method of degradation of one carbon at a time, it is possible to distinguish between direct uptake of injected fatty acids and synthesis of a labeled compound via β-oxidation of the injected material followed by resynthesis. Thus, by determining the ratio of activity in the carboxyl carbon to that of the whole fatty acid (percentage relative carboxyl activity) one can distinguish direct uptake of injected fatty acid from its prior oxidation to acetyl-CoA and reutilization of this radioactive com-

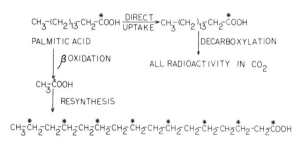

Figure 10A. Distribution of radioactivity in palmitic acid following direct uptake and β-oxidation and resynthesis. Alternate carbons are labeled. Decarboxylation should give CO_2 that contains only one-eighth (12.5%) of total activity.

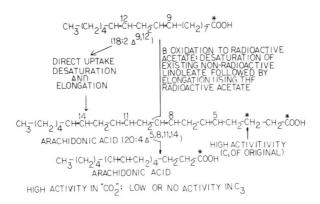

Figure 10B. Distribution of radioactivity in arachidonic acid following direct uptake and β-oxidation and chain elongation of labeled linoleate.

ponent. Using these methods, Dhopeshwarkar *et al.*, (1973 and references therein) have clearly established that fatty acids, both essential and nonessential, can be taken up by the brain from the circulating blood even by mature adult animals (Figs. 10A and 10B).

4.3. NONMORPHOLOGICAL NATURE OF THE BLOOD–BRAIN BARRIER SYSTEM

There are several nonmorphological explanations that have been offered to explain the effects of the blood–brain barrier. One such explanation deals with the principle of mediated transport. A carrier substance (usually a protein) facilitates transport of water-soluble substances across the membrane. Thus, the difference between transport of glucose and fructose could very well be the difference between the carriers of these two substances. The availability (concentration) and properties (saturating capacity) of the carrier substances may be different and hence there may be a large difference between the transport of any two substances. A further complication is the competition between different substances for the carrier proteins that can regulate uptake from blood into the brain. Another explanation that has been offered relates to the degree of saturation or metabolic requirement of any substance in question. Dobbing (1961) argues that the uptake of a substance by a tissue is influenced by the availability of suitable metabolic activity of that substance. The metabolic pathways and enzyme systems in the brain may not be comparable in all respects with other metabolically active tissues. Therefore, restricted entry may be explained on the basis of metabolic potentialities of the brain for that substance.

The brain has several compartments, some are large with a slower turnover rate (myelin), whereas others are turning over rapidly. Thus, in mature animals insofar as myelin is concerned, there is neither a rapid uptake nor exit. In the younger animals myelination is progressing rapidly and, therefore, both synethetic and transport activities are proceeding at a rapid rate, but once myelination is complete this process slows down. This is reflected by the difference in uptake of substances needed for myelination in the younger versus the older animals. The progressive development of the barrier in the pre- and postnatal period depends on the species and is not uniform in all mammals. Human fetuses exhibited a restricted passage of trypan blue but in rabbits, weighing 4 to 22 g, there was no such restriction (Davson, 1967). In rats, the barrier seemed to have developed in the fetus as early as midway through the period of gestation (Millen and Hess, 1958).

Another idea to support the hypothesis that the barrier is a non-morphological phenomenon comes from the work of Owman and Rosengren (1967). These workers proposed that the barrier is enzymatic in nature, and, using histochemical techniques, they show that GABA and L-dehydroxyphenylalanine (L-DOPA) are degraded within the capillary endothelium. Further, only when this degradative capacity is overcome can these compounds get into the brain from the circulating blood. Other explanations, such as the sink action, also have been proposed (Levine and Scicli, 1969).

From the above discussion, it is quite apparent that the phenomenon of restricted uptake or transport of substances from the blood into the brain observed in many experiments is unique to the CNS. Beyond the recognition of the fact that incorporation of a compound is a function of both barrier and metabolism of the tissue, we do not know the exact mechanism of its operation or its physiological nature. However, its usefulness is readily understandable. Some of the toxins that are introduced into the blood cannot permeate into many areas of the brain, thus this restriction can offer protection from harmful effects. Other regions of the brain, such as area postrema, do not exhibit such a restricted permeability. These areas are close to the "vomiting center" in the brain; the ingestion of a toxin can induce vomiting thus helping in excretion of the toxin.

Another use of this phenomenon is realized in nuclear medicine. Following an intravenous injection, ^{131}I-labeled serum albumin usually does not leak out of the capillaries in the brain but will do so in areas damaged by a tumor. This will show up in the radioscan helping the clinician in diagnosis. Sophisticated imaging instruments computerized to a great extent can now pinpoint the damaged areas to help in treatment, sometimes without invasive procedures. Newer techniques of temporary "opening up" of the barrier have far-reaching possibilities in enzyme replacement therapy and treatment of brain tumors.

The importance of the blood–brain relationship to nutrition is obvious. Unless uptake of compounds, dietary or therapeutic, is rigorously examined, they would have no effect on the brain physiology or function. Similarly, to achieve maximum benefits, a dietary regimen needs to be instituted very early when the uptake of many nutrients is also high.

5

ALCOHOL: EFFECTS ON THE CENTRAL NERVOUS SYSTEM

5.1. EFFECTS ON THE ADULT

Alcoholism has become a public health problem. The Surgeon General reports (1979) that alcoholism is increasing in men and women of all ages and that the seriousness of the problem is increased severalfold when statistics show that 20–25% (about 3 million) of 14- to 17-year-old adolescents are problem drinkers, i.e., intoxicated at least once per month. Thus, the effects of alcohol on health have received much attention in numerous disciplines—clinical, behavioral, biochemical, nutritional, and others. In this chapter, discussion will be restricted to the effects of alcohol consumption on the central nervous system in adults as well as fetuses.

Two aspects need to be discussed with regard to alcohol and its relation to CNS function; one, how ethanol affects brain function via direct pharmacological effects, and, second, how ethanol affects the bioavailability of other nutrients, such as vitamins, amino acids, and minerals, that are vital for brain physiology. In dealing with the second aspect, it should be pointed out that oxidation of ethanol (7.1 cal/g) is almost complete in the human body, i.e., the metabolic intermediates are not channeled into synthesis of other compounds such as amino acids or fatty acids. This means that alcohol ingested in the form of beer or wine does not contribute to the buildup of any nutrients and is used

41

solely for energy production. In fact, by directly interfering with gastric emptying (Barboriak and Meade, 1970), acid secretion in the stomach (Davis *et al.*, 1965), and intestinal motility and subsequent diarrhea (Robles *et al.*, 1974), alcohol can certainly have an overall effect on the nutritional status. The interference by alcohol in the normal process of absorption of nutrients has been shown in the case of methionine and other amino acids (Isreal *et al.*, 1968). Since methionine in the form of S-adenosylmethionine (SAM) is a methyl group donor, lack of this amino acid may have a direct effect on O-methylation of neurotransmitters. A direct effect on the high-affinity active transport of vitamins (thiamine in particular) in the G.I. tract leading to malabsorption has been reported (Hoyumpa *et al.*, 1974). Turning to other nutrients such as essential minerals, Ca^{2+} for example, it was found that in chronic alcoholism calcium levels were decreased in blood plasma (Ogata *et al.*, 1968); the correlation between decreased blood levels and the amount in the brain has not been established.

The metabolism of ethanol by the liver indirectly affects brain function. The oxidation of ethanol in the liver is a two-step reaction: ethanol → acetaldehyde → acetate → CO_2 + H_2O. In the first step ethanol is converted to acetate and in the second step acetate is completely oxidized to CO_2. Although the liver has the capacity to catalyze both steps, the acetate produced in the liver is secreted into the bloodstream for transport to other organs where complete oxidation to CO_2 can be achieved (Lundquist, 1962; Forsander and Raiha, 1960). The energy demand seemed to be satisfied by the first step and the metabolites released by the liver into the circulation could then exert their influence on the CNS. Chronic consumption of alcohol may lead to overloading of the oxidative capacity of the liver (up to a certain point the liver has a built-in capacity to increase enzyme activity) and thus may increase aldehyde levels in the circulation. Competitive inhibition of the oxidative metabolism of aldehydes, derived from neurotransmitters, by acetaldehyde, derived from ethanol, can interfere with normal metabolic pathways (Lahti and Majchrowicz, 1967; Dietrich and Erwin, 1975) (Fig. 11).

The competition between acetaldehyde and aldehyde intermediates of neurotransmitter metabolism as well as other important reactions require further scrutiny. For example, consider the reactions in Figure 12. These condensation products, tetrahydroisoquinoline (TIQ) alkaloids, behave like false transmitters (Heikkila *et al.*, 1971) whose affinity was found to be much lower. These compounds have also been shown to inhibit O-methyltransferase activity (Collins *et al.*, 1973). A more direct effect of ethanol is its synergistic effect with hypnotics on sleep time (Blum *et al.*, 1971). However, the presence of low levels of acetaldehyde

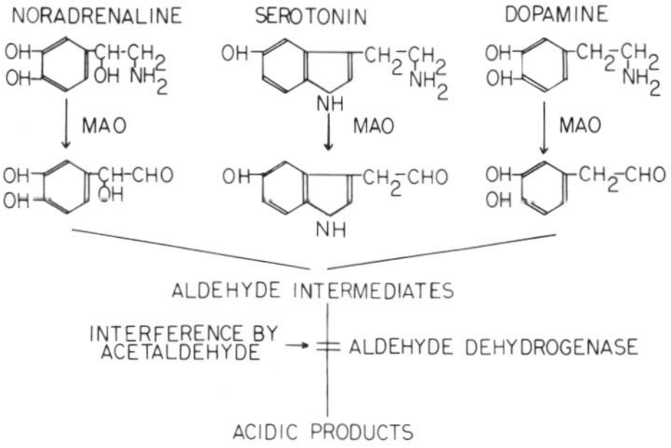

Figure 11.   Competitive inhibition by acetaldehyde on the oxidative metabolism of aldehydes derived from neurotransmitters.

in the brain following a single dose of ethanol should be taken into account when evaluating the above results. In this respect, some early work by Beer and Quastel (1958) had indicated that acetaldehyde at very low levels brought about a marked inhibition of potassium-stimulated respiration of a brain cortex slice. There was no increase in inhibition with an increase in concentration of aldehydes, unlike the alcohols. Truitt *et al.*, (1956) found that acetaldehyde produced a limited degree of uncoupling of oxidative phosphorylation but corresponding alcohols did not.

5.1.1.   Effect of Alcohol on RNA and Protein Synthesis

Tewari and Noble (1975) have reported that when a 10% ethanol solution was ingested for a long time, there was a marked decrease in the synthesis of nuclear as well as mitochondrial RNA. A separate study

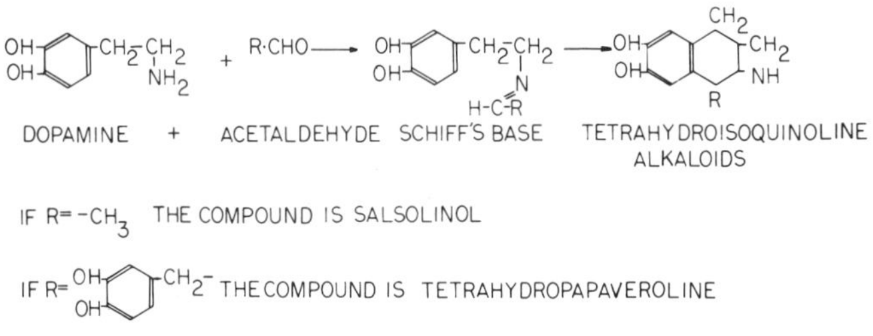

Figure 12.   Formation of TIQ alkaloids which behave like false transmitters.

by the same workers (Noble and Tewari, 1973) showed that as a result of chronic ingestion of alcohol, even the synthesis of protein decreases in the brain. However, one needs either a higher dose or a chronic ingestion of ethanol to bring this about. The effect on protein synthesis could be due to dissociation of ribosomal subunits (Tewari and Noble, 1977).

5.1.2. Effect of Alcohol on Cellular Membranes

Chin and Goldstein (1977) have claimed that a change in fluidity of brain membranes follows alcohol consumption. The increase in fluidity of the membrane observed by the electroparamagnetic resonance signal seemed to be small. Further study by the same authors (Chin *et al.*, 1979) showed that in ethanol-tolerant mice there was an increased amount of cholesterol in the brain membranes.

It should be remembered, however, that it is very difficult to single out the direct effects of ethanol on the nervous system. Many of the effects appear to be similar when compared to other hypnotic or tranquilizing drugs. The exceptions to these are the inhibitory effects of alcohol on the release of oxytocin, vasopressin, and other hypothalamic peptides (Kalant, 1975). Chronic ingestion of alcohol is usually accompanied by nutritional deficiencies and these will be discussed separately under appropriate sections.

5.2. EFFECTS OF ALCOHOL ON THE FETUS: FETAL ALCOHOL SYNDROME (FAS)

Some general considerations regarding the growth of the fetus can be briefly reviewed before discussing the effects of alcohol. Recently introduced techniques, like ultrasound scanning, have made it possible to measure fetal growth in humans (Campbell, 1974). From the studies by Usher and McLean (1974), it seems that the most rapid growth of the fetus occurs between 14 and 34 weeks when the fetal weight increases from about 50 g to approximately 2500 g and the average weight at term is about 3300 g (7–8 lb). These increases in weight and size seem to slow down during the last period of gestation just before birth. In any case, the rapid increase needs a constant supply of oxygen and nutrients provided by the maternal circulation. The growth of the fetal brain occurs rapidly and, therefore, the ratio of brain weight to body weight is higher in the fetus than in adults.

As mentioned before, the sole source of nutrition for the fetus is maternal blood. It is well known that the maternal blood supply does not mix with the fetal blood but is separated by the placenta. Assuming that the mother is well nourished, the fetus will derive all its needs from this source. The placental transport (somewhat similar to the blood–brain transport) is also selective. The transport could be by simple diffusion from higher to lower concentration and does not require any expenditure of energy. The diffusion of all substances is not equal; whereas lipid solubility may be an asset, high molecular weight is a hindrance. A more important method of transport is represented by "mediated transport." In this process the substances combine very strongly with a carrier molecule (mostly protein in nature) which carries them across the membrane and liberates them on the other side. In the case of active transport, when molecules have to be transported against a concentration gradient, energy is needed and usually provided by the metabolic reactions occurring in the cell. A carrier-mediated "piggyback ride" is common for glucose, amino acids, and fatty acids. The movement across the membrane of some cells of glucose and amino acids occurs only in association with N^+ transport. The carrier proteins are very selective in binding; for example, the binding of D-glucose is very strong in comparison with L-glucose. Finally, receptor binding proteins are important in high-affinity transport systems. Some of these exhibit substrate-induced conformational changes in the translocation step of transport. For larger molecules like proteins or polypeptides, pinocytosis is the preferred form of transport. In this process, the plasma membrane invaginates and engulfs the substance creating vacuoles that pass the large molecules to the opposite side.

The occurrence of FAS is increasing because the incidence of alcohol dependency is increasing in women, even in those of child-bearing age. What is discussed below is due solely to alcohol intake and not complicated by other nutritional deficiencies. Alcohol has no restriction in passing from the maternal blood into the fetus and the concentration in the two compartments, which remain separate and do not mix, is almost equal (Weathersbee and Lodge, 1978). Children born to alcoholic mothers display facial characteristics and growth deficiencies that can be diagnosed with relative ease (Clarren and Smith, 1978). These include smaller head circumference, low nasal bridge, short nose, and thin upper lip. It should be pointed out here that these characteristic signs in appearance are common to all racial groups. Usually these children are deficient in adipose fat and in overall weight. Table 2 describes the features and manifestations of FAS. The features described are all physical handicaps, but several workers have reported mental disturbances as well. The average IQ of children with FAS is around 68, considered

Table 2. Principal Features of Fetal Alcohol Syndrome[a]

Feature	Manifestation
Central nervous system dysfunction	
Intellectual	Mild to moderate mental retardation*
Neurologic	Microcephaly;* poor coordination and hypotonia†
Behavioral	Irritability in infancy;* hyperactivity in childhood†
Growth deficiency	
Prenatal	Less than 2 standard deviations for length and weight*
Postnatal	Less than 2 standard deviations for length and weight;* disproportionately diminished adipose tissue†
Facial characteristics	
Eyes	Short palpebral fissures*
Nose	Short, upturned†; hypoplastic philtrum*
Maxilla	Hypoplastic†
Mouth	Thinned upper vermillion;* retrognathia in infancy;* micrognathia or relative prognathia in adolescence†

[a] Reproduced from Clarren and Smith (1978).
* Feature seen in > 80 percent of patients.
† Feature seen in > 50 percent of patients.

mildly retarded (Streissguth *et al.*, 1978). FAS causes deep-rooted permanent damage to the brain; restoration with balanced diet or rehabilitation in a home with excellent child care cannot restore the mental handicaps or low IQ scores. Additional factors such as heavy smoking or use of drugs make matters worse. The drugs that need special attention are cancer chemotherapeutic drugs, anticoagulants, anticonvulsants (diphenylhydantoin), oral antidiuretic agents, and some very commonly ingested compounds like caffeine and aspirin. The question about upper and lower limits of alcohol intake cannot be answered because no one knows the lower limit. However, with more than 2 oz. of absolute alcohol per day (equal to or more than four drinks per day), Hanson *et al.* (1978) report an incidence of 19% born with signs of FAS. With fewer than two drinks per day the percentage of affected children decreased to 2%. However, the possibility of lack of gross abnormalities but presence of subtle abnormalities with low intake may not be ruled out and the risk would not be worth a short-lived pleasure!

Consumption of alcohol affects intrauterine growth and this results in the birth of "small-for-date" babies, sometimes referred to as small for gestational age (SGA). Decreased birth weight is accompanied by smaller head circumference and body length (Streissguth *et al.*, 1980). Alcohol consumption has also been reported to induce stillbirths and in combination with smoking the incidence increased two to five times, depending on the amount of alcohol intake and degree of smoking. Samples of brain after autopsy studied by Clarren *et al.* (1978) show not only a decrease in size of the brain from infants born to mothers addicted to alcohol but also that the pattern appeared altered, including an abnormal cell migration (Fig. 13).

Since milk is the only source of nutritional need in the newborn, suckling behavior (i.e., suckling pressure which would control milk intake) is an important parameter. Martin *et al.* (1978) have reported that infants born to mothers who consumed alcohol during pregnancy had significantly weaker suckling pressure.

Many other biochemical abnormalities are probably occurring in babies born with FAS, but for obvious reasons one has to rely on autopsy material, and many times this is unsuitable for certain biochemical investigations. There is, therefore, a great need for animal studies.

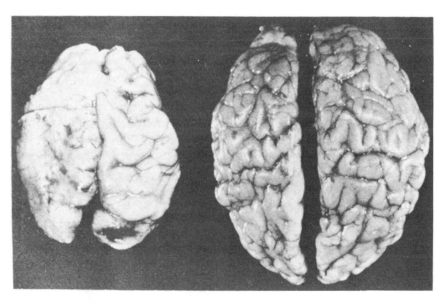

Figure 13. Comparison of normal newborn brain (right) with brain of FAS infant. Note size and gyral pattern differences. Reproduced from Streissguth *et al.* (1980).

5.2.1. Animal Studies

Animal studies offer other advantages, such as the ability to strictly regulate and monitor alcohol intake and timing of intake in relation to the period of gestation. Finally, the frequency of intake can also be regulated. Many laboratory species, e.g., rats, mice, guinea pigs, and dogs have been used in such studies. In general, a variety of defects have been reported, e.g., eye defects and cardiovascular and head malformations, in mice with a large intake of alcohol, and delayed ossification with smaller doses of alcohol ingestion in rats (Streissguth *et al.*, 1980). Biochemical parameters such as decreased protein synthesis (Rawat, 1975), delayed myelination, and hormonal disturbance (Rosman et al., 1975) have been reported in the literature. Detering *et al.* (1980) have reported that dopamine levels were significantly lower in fetuses of dams consuming 35% of the calories in the form of alcohol. The levels remained lower even after birth. Similarly, NA levels were also reduced in the fetuses and in the pups after birth. The hypothalamic region of the brain was found to have a 30–60% decrease in the level of NA. This clearly indicated the damaging effect of alcohol ingestion during pregnancy on the NA neurons of the fetus. Whether this can be reversed by abstaining from alcohol in later life is not yet clear. Hyperactivity, sometimes attributed to intake of food additives, has been found in pups born to dams fed alcohol (Branchey and Friedhoff, 1976). Since pair-fed animals were used in most studies, one can conclude that it is the ingestion of alcohol, rather than the nutritional deficits that usually accompany alcoholism, that cause morphological, physiological, and biochemical abnormalities.

Apart from the serious effect on the fetus and newborn, alcoholism in adults leads to nutritional deficits. A far more dangerous outcome, however, is death on the highway by car-related accidents, which has reached epidemic proportions in certain countries.

EFFECTS OF MALNUTRITION ON BRAIN DEVELOPMENT

6.1.   CALORIC INSUFFICIENCY (HUMAN STUDIES)

Human experience indicates that conditions of insufficient food, hunger, and starvation occur quite commonly in many parts of the world. The United States has been spared devastating famines or war damages for a long time and enjoys the best agriculture and an ample food supply. However, this has not been the case in Europe, devastated by two world wars in recent history, or Africa and Asia which have faced many acute food shortages, sometimes due to natural calamities and sometimes man-made. As mentioned earlier, the impact of nutritional deficiencies is felt most in infants and children because their nervous systems are still developing, whereas, in adults the brain is usually spared the effects of malnutrition. The war years in Europe and the eruption of famine conditions in other countries have provided the bases for evaluating the effects of caloric insufficiency [and in some cases protein–calorie malnutrition (PCM), and a lesser degree of essential fatty acid (EFA) deficiency] on some parameters of brain damage, behavioral change, and intelligence (IQ tests).

The brain "growth spurt" in humans covers the prenatal (last trimester of fetal life) and the postnatal period of 18 months. Thus, most of the studies in humans have been concentrated around this period of growth and development. In all cases, it is important to know the severity of nutritional deprivation and the period of duration. Thus, severe famine conditions resulting in a wide prevalence of marasmus or severe caloric deficit offered a setting to evaluate the effects of such malnutrition on the CNS.

Caloric insufficiency during pregnancy and lactation has led to marasmic infants characterized by an apparent large head relative to a very thin, small body, with arms and legs so emaciated that they appear to be mere bones, and swollen knee joints. The seriousness of this condition can be realized from the fact that marasmic children without any remedial effort just cease to grow. Apathy and hyperirritability are also noted in marasmic infants, indicating CNS involvement. Since the adult has reached normal height, the effects of starvation can no longer influence growth, but in a child deprived of calories, growth to normal height is permanently stunted. Brown (1965) has reported that, in general, the brain weight of children who died of marasmus was greatly reduced. Similar studies have been reported from South America (Winick and Rosso, 1969). Studies on the brain weight of infants considered "small for date" (mostly due to maternal malnutrition rather than genetic abnormalities) from India indicated that not only was the brain weight significantly reduced, but different parts of the brain were affected differently. For example, on a percentage basis, cerebellum and medulla oblongata were affected more than cerebrum but no significant differences in biochemical investigations, such as in amount of proteins and lipids were found (Sarma and Rao, 1974). Malnutrition has been identified as the major cause of reduced head circumference in marasmic children. Measurement of head circumference, being a noninvasive parameter of brain size and weight, is used all over the world. The earlier the nutritional deficit, the more severe the decrease in head circumference. This assumption was not only proved to be true in most cases but was further strengthened by comparing marasmic children with those suffering from kwashiorkor. Marasmus is usually noted in infants much earlier than kwashiorkor and so the early nutritional stress affects the brain size much more than kwashiorkor. (It should be pointed out here that the two conditions do not have the same nutritional etiology and are, therefore, different; here emphasis was placed only on the timing of nutritional stress.) Thus, the impact of marasmus on brain growth is more severe.

In the case of PCM, one usually finds a low serum albumin level and sometimes hypoglycemia. The nutritional deficiencies can actually get worse with diarrhea (not uncommon under these conditions). Resistance to infection is reduced considerably and all these conditions together lead to the high mortality seen in certain areas of the world, for instance, in Biafra. A combination of all these factors, needless to say, affects all parameters of brain development. The malnutrition of the infant actually begins *in utero*. For example, approximately 4 kg of subcutaneous fat depositions in the mother during pregnancy can be a major source of calories and fatty acids during lactation. If malnutrition occurs during pregnancy, due to some unavoidable economic conditions or avoidable dieting, then lactation can turn out to be inadequate. In fact, Jelliffe and Jelliffe (1979) point out that, for the newborn up to about 9 months, the breast functions for the infant like the placenta does for the fetus. These authors point out that the differences in human milk and cow's milk using modern analytical techniques show far more diversity than differences in fat content or macroconstituents (Fomon, 1974). Also the nutritive value of breast milk is not always the same at all times. The early morning sample is always higher in fat. Since the diet of the mother has a great influence on the composition of milk, there is a possibility of seasonal variation depending upon abundance of seasonal commodities in the diet (i.e., fruits, vegetables). The fatty acid composition of the milk is also influenced by the type of dietary fat during lactation. Whether breast milk can supply all the needs of the infant even when the mother is consuming an adequate diet is still debatable. In this connection, early work (Welbourn, 1955; Jelliffe, 1954) showed that the growth rate of babies solely on breast milk was satisfactory up to the first 5–6 months. However, recent studies have reduced the age to 4 months only (Bailey, 1965). In one study done in Pakistan (Lindbad *et al.*, 1977), the growth curves of babies fed unsupplemented breast milk were comparable to Western standards only up to 3 months. Thus, it seems that breast-feeding up to 3–4 months will assure an adequate supply of nutrition to the infant and also protect it from infection. This last point may not be trivial in some areas where there is lack of sanitation and the clean water supply is well below standard. Thus, formula milk made from contaminated water may be more harmful than the nutritional inadequacies it is supposed to eliminate. In any case, supplemental foods need to be introduced after the initial 3–4 months to ensure adequate levels of calories and essential nutrients.

The incidence of small-for-date babies should not be confused with babies born prematurely and thus weighing much less than full-term babies. The former example indicates intrauterine growth retardation due mostly to inadequate diet (examples of excessive intake of alcohol and/or smoking may also lead to intrauterine growth retardation as mentioned earlier). Small-for-date babies many times have feeding difficulties, such as the inability to suck, inhalation of food, and vomiting. They may also have nutritional deficiencies (depending on the maternal diet) of vitamins K and E or folic acid; another deficiency commonly noticed is iron deficiency. Metabolic disorders such as impaired thermoregulation due to decreased adipose fat, neonatal hypoglycemia due to increased glycogen stores, neonatal hypocalcemia, and protein (milk) intolerance (Roe, 1979) have also been noted. Neonatal hypoglycemia, if untreated, may eventually lead to seizure, cerebral palsy, and mental retardation (Haworth and Vidyasagar, 1971). Dobbing and Smart (1973) have listed several lasting effects of intrauterine growth retardation which include small (but not uniformly small) brain size, i.e., cerebellum smaller than the rest of the brain. There are fewer cells, as determined by the DNA content, and once again the cerebellum, in particular, has fewer late-dividing granular neurons. The lipid content is decreased even when corrected for the smaller size. The activity of enzymes such as acetylcholine esterase per unit of fresh weight was found to be surprisingly higher in previously undernourished animals than in controls.

The small-for-date babies arising from maternal malnutrition exhibit growth failure of most organs including the brain. In contrast, a decreased blood supply to the infant affected the liver much more than the brain (Winick et al., 1972a). For example, there was a 50% decrease in liver DNA but no such reduction in the brain cells. This depletion of liver cells can cause several metabolic defects, including hypoglycemia, perhaps due to depleted glycogen stores. At least in experimental animals, such as monkeys, placental insufficiency (produced by ablation of part of the placenta) caused a reduction in cell number, RNA, and also protein-per-cell in the liver and skeletal muscle (Hill et al., 1971). Infant mortality goes hand-in-hand with low birth weight, and among survivors the chances of mental retardation are relatively high; however, this does not mean that every low-birth-weight baby is going to have either mental retardation or some other kind of severe abnormality.

As mentioned before, the human situation insofar as effect of malnutrition is concerned is very complex, since socioeconomic factors also play a very important role. The impact of such variables has already been emphasized in the discussion of FAS. Several relatively large-scale studies have been conducted in parts of the world where adverse nutritional situations occur, due either to natural disasters like drought and consequent severe famine or to man-made problems arising from political rivalries, blockades, and wars. Certain other situations unfortunately arise out of poverty, and not enough food supply to satisfy a modest daily requirement. Cebak and Najdanvic (1965) reported a significantly lower IQ in children suffering from severe caloric deficiency as compared to other children receiving normal, adequate nutrition. The authors emphasize the time of nutritional stress. The earlier the stress, the greater was the aftereffect. Similarly, Liang *et al.* (1967) observed that many Indonesian children in their teens who suffered from early malnutrition scored poorly when compared to well-nourished tall children. Some behavioral scientists have complained that the tests to study IQ or performance are not suitable for all children due to different cultural, social, and religious backgrounds. To overcome this objection in a study in India, children were selected from similar backgrounds, such as educational background of parents, religion, and socioeconomic status in addition to normal factors such as age and sex. As in other studies, the Indian study also concluded that nutrition was the deciding factor (Champakam *et al.*, 1968). One of the differences between these studies and earlier studies is that the children tested in the Indian study were diagnosed as suffering from kwashiorkor rather than marasmus. One can also mention here that if the children used in these studies include those who were hospitalized, this period of separation from parents and relative inactivity should be taken into account. Obviously human studies are difficult to control and there are some factors that cannot be controlled easily. Thus, comparison of the IQ of these children with "normal" children cannot be without some reservations.

Cravioto and co-workers (1966) have also done similar studies in Guatemala and Mexico. They found that children with short stature (body weight and height) caused by malnutrition showed a convincing lag in development of sensory integration examined by such tests as placing blocks in openings and differentiating block shapes by touch. The stunted growth, shorter height, and lower weight were not due to similar stature of the parents; many other variables such as cash income and portion of income spent on food, were also ruled out. The only cause of poor performance was malnutrition during the period of rapid growth and development of the CNS.

From these studies one tends to conclude that malnutrition has a serious deleterious effect on the normal growth and development pattern. However, a study done in Holland seems to disagree with this conclusion (Stein *et al.*, 1972). During the war years, 1944–1945, parts of the Netherlands suffered extreme famine conditions. The average intake of food constituted a bare caloric intake of about 750 or less. Yet another part of the same country offered approximately double the amount per day. This unique situation presented a model to study the growth and performance of children born during this time in the two regions. To the surprise of everyone in this area of investigation, it was found that there was no significant difference in the performance in the two groups of children. The tests were, however, conducted many years later, i.e., the gap between exposure to malnutrition and the time of the tests was very long. The birth weights of the children from the two areas were different and so it was even more of a surprise that no effect on mental development could be detected. Earlier it was mentioned that small-for-date babies had many disadvantages and failed to "catch up."* The Netherlands is one of the countries which for years had enjoyed a higher standard of living, and so, other than lack of calories during a relatively brief period, the expecting mothers and their children did not have to face nonnutritional deprivations like children in other poorer countries; this may be one of the many possible causes of the lack of correlation between malnutrition and mental performance.

Malnutrition need not always be due to lack of food. It can be due to an imbalanced diet (e.g., lack of macro- or micronutrients or excess of one component in the presence of marginal amounts of an essential component). But although much work along these lines has been done, it has been restricted to animal studies, to be discussed later in this chapter. Malabsorption due to known causes like intestinal pathology or necessary medical intervention (e.g., surgical removal of part of the gastrointestinal tract or X-ray treatment) does occur in a certain number of children. In other cases, it could be idiopathic or from unknown causes. It could also be due to pancreatic insufficiency or fatty diarrhea caused by diseases like cystic fibrosis or tropical sprue. When studies were done to compare these children with control siblings, it was found that the mental retardation and other impairments occurred only if the malnutrition period started very early in life, i.e., below 5 years of age (Lloyd-Still et al., 1972).

Autopsies of children who died of established nutritional deficiencies have also pointed out that total protein, total RNA, total DNA, total cholesterol, and total phospholipids are all reduced (Rosso et al., 1970).

The question of recovery from deleterious effects of malnutrition by remedial balanced nutrition in humans is not easy to answer. A summary of various findings will be discussed later.

6.2. ANIMAL STUDIES

Animal studies of the effect of malnutrition on the brain have been carried out by many laboratories including our own. The importance of

*The author had a chance to meet a woman (now a practicing physician in the United States) who lived in Holland during these adverse times and she related that although the general population ate only 750 cal, the expecting mothers always managed to eat more. This may be the reason why such alleged malnutrition did not affect mental performance in children.

the subject has received worldwide attention. Dobbing and Davison in England, Galli and Paoletti in Italy, Svennerholm in Sweden, Winick, Zamenhof, and Chase in the United States, and many other groups active in this field have collectively contributed many parameters which, for obvious reasons, cannot be studied in human populations.

The method of producing caloric insufficiency in laboratory animals is to cull the number of pups suckling per mother. Thus, 16–18 suckling pups per mother provides a malnourished group whereas 6–8 provides a good balanced control group. If overfeeding is desired, only 2 or 3 babies can be left with the lactating mother rat. In the latter group, it is not known whether the amount of milk secreted by the lactating mother rat also decreases rapidly; thus, the pups may not have an oversupply but just a normal, adequate supply. This method of producing malnutrition is obviously restricted to caloric insufficiency. However, other methods of reducing the milk supply by an imbalanced diet have also been tried and will be mentioned later on.

Widdowson and McCance (1960), over 20 years ago, using this method of unequal litter size, found that malnutrition of this type results in decreased brain weights. The decrease was higher than the difference in body weights. This observation had been confirmed in many later studies by Dobbing (1964), Winick and Noble (1966), and de Guglielmone et al. (1974). A similar result was noted when the nursing time was restricted rather than varying the total number of suckling pups per mother (Rajalakshmi et al., 1967). Another method of reducing milk supply is to decrease the amount of protein in the diet fed to pregnant and lactating dams (Sobotka et al., 1974). Irrespective of the method used, caloric deprivation resulted in reduced brain weight compared to controls. Species other than the rat, pigs or monkeys, for example, also showed similar effects (Dickerson et al., 1967; Kerr et al., 1973).

Next to evaluating size and weight of the body (ratio of brain to body weight may not change because both are reduced in newborns and suckling pups), the most widely studied indicator of brain "damage" has been the cell number. Since DNA is constant within the diploid cell of all species and most of the cells in the brain are diploid, measuring DNA of the whole brain gives, by simple calculation, the total number of cells in the brain. One should always bear in mind that DNA determination cannot differentiate between types of cells and that there is a migration of the cells even after all cell division ceases. Thus, an increase in DNA could be due not to net increase in number of cells in that region in situ but to migration from another region.

Winick and Noble (1966) have used this parameter extensively in their studies on the effect of malnutrition on the brain. Their study shows that brain weight as well as protein content of the whole brain was reduced. DNA or cell number was reduced by 34% when measured at weaning (21 days). Swaiman *et al.* (1970) also found that a reduced caloric intake resulted in a 30% reduction in DNA at 7 days, 25% at 14 days, and 10% at 21 days. Thus, the magnitude of decrease might vary with experiments in different laboratories but the conclusion has always been the same.

The time of nutritional imbalance during critical periods of growth is yet another very important factor. For example, Dobbing (1972) in his studies has chosen the following conditions: Four groups of animals (1) G^+L^+, (2) G^-L^+, (3) G^+L^-, and (4) G^-L^-, where G stands for gestational period, L for lactational period, and $^+$ or $^-$ designates good or poor nutrition. When body weights of these animals were compared, he found that G^-L^- were most severely affected as expected but surprisingly G^+L^- were affected almost equally. On the other hand, G^-L^+ were very much like G^+L^+. So his conclusion was that nutrition during the lactation period had far-reaching effects but restriction during the gestational period had very little, if any, effect. Further, keeping the litter size uniform, pups from mothers underfed during gestation or lactation were cross-fostered at birth. Yet the same results were obtained, i.e., L^-; malnutrition during lactation seemed to be the deciding factor. It is interesting to note from these studies that nutritional manipulation during the "brain growth spurt" (between 5 and 20 days postnatally) affected body weight. Other effects of caloric restriction observed by Dobbing and co-workers (Dobbing, 1964, 1971, 1972; Dobbing and Sands, 1973; Dobbing and Smart, 1973) include ultimate microcephaly, permanent reduction in total cell number of the whole brain, reduction in brain lipids (in particular myelin lipids), and a distortion of the enzyme pattern in the brain. The only debatable question that remains to be decided is whether to include the rapid neuronal growth phase in the vulnerable period. It seems that this question is still being debated and remains unresolved.

Winick and his co-workers (Winick *et al.*, 1972a,b) have done exhaustive studies on the effect of malnutrition on cell number and rate of cell division. They report that caloric deprivation during the growth spurt will retard cell division which results in a decrease in the total cell number. The rate of cell division is not uniform throughout the brain, e.g., the rate is higher in the cerebellum than in the forebrain. A good parameter of cell division is the rate of DNA synthesis via DNA polymerase activity. Thus, measuring DNA polymerase activity in various parts of the brain at different time periods can be used to estimate the changes in cell division. In the cerebellum there are two peaks of DNA synthesis, one at 7 and the other at 13 days. Malnutrition was found to affect this important enzyme; the activity of the DNA polymerase was very much reduced in the brain of rats reared on insufficient calories. In addition to this, in general, protein synthesis was also reduced as a result of undernutrition (Miller, 1970). Winick explains this as an outcome of desegregation of polysomes which do not synthesize proteins as well as intact polysomes.

Dobbing and Sands (1973) have demonstrated that in the human brain there are two peaks of cell division, one from 15 to 20 weeks of gestation and the second beginning at 25 weeks and continuing until birth. Thus, if the first peak of cell division denotes neuronal proliferation, malnutrition during this stage will affect the neuronal cell number, whereas the second peak due to glial proliferation will affect myelination. Thus, there is a parallel between studies done with animal models and findings of human autopsies performed in cases of death due to malnutrition.

Besides the debate about which time period is most critical, neuronal or glial proliferation peak, other more serious questions remain unanswered. For example, it is not clear whether cell number, cell size or extension and elongation of the neuronal processes, branching of dendrites, or the establishment of synaptic connections collectively affect brain function or whether any one of the above is known at this time. Thus, even if caloric undernutrition affects most of these events we still are far from a clear understanding of the relation between brain function and early undernutrition.

6.3.   EFFECT OF PROTEIN DEFICIENCY

Apart from a general caloric deficiency, other deficiencies also have proved to affect brain growth. Mention has already been made of the use of a low-protein diet that led to decreased lactation. Apart from this effect, Zamenhof and co-workers (1968) very early in their studies established that female rats maintained on a low-protein diet (8%) for 1 month before mating and during the gestational period yielded affected cerebral hemisphere size and containing 10% less DNA and 19% less protein compared to controls. This indicates that the cell number was reduced at birth and since, in the rat, this period is related to neuronal proliferation the neuronal cell number was decreased. In their next experiment (Zamenhof et al., 1972) offspring (F_1) of mothers (F_1) on a low-protein diet were mated with normal males and females. Their offsprings (F_2) received a normal diet from birth. Body weight, cerebral weight, and protein were determined at birth and at 30 and 90 days. The pre-

and postnatal (F_1) group showed a 7% decrease in cerebral DNA at birth. Cerebral DNA was decreased significantly (6%) in the F_2 offspring at birth. The conclusions drawn from these studies were that nutritional insults sustained by females over one or two generations were expressed in their offspring. However, the deficiencies of the F_2 generation were reversed by dietary means. The decreased amount of dietary protein used in these studies and the omission of one essential amino acid in the diet seemed to equally affect prenatal brain development. Zamenhof *et al.* (1974a) fed a diet in which a complete chemically defined mixture of L-amino acids replaced the intact protein. Another group was fed the diet containing all the above amino acids except one essential amino acid, tryptophan, lysine, or methionine, during 0–21 or 10–21 days of pregnancy. The results showed that even when all amino acids were included in the diet either all through pregnancy or during the last half of pregnancy fetal growth was significantly inferior when compared to those fed the pelleted stock diet. The significant differences were restricted to body weight; cerebral parameters such as weight and DNA (cell number) and placental parameters (weight, DNA) were not affected. Thus, it seems that present knowledge of a faultless amino acid mixture for optimal fetal development is incomplete. The omission of either tryptophan, lysine, or methionine resulted in brain growth deficiency similar and comparable to protein-deficient diets. Therefore, the authors conclude that omitting one essential amino acid was comparable to total absence of dietary protein. These results are in agreement in some respects with those of Niiyama *et al.* (1970) who found that omission of one essential amino acid caused reproductive failures and lower neonatal weights, but when steroids were given, pregnancy was maintained. However, these studies did not include examination of brain growth parameters.

Zeman and Stanbrough (1969) have also shown that even if the low-protein diet (6% casein) was started on Day 0 of pregnancy, fetal weight was decreased by the 16th day. The total DNA was significantly decreased in the livers of 18-day fetuses and all tissues (including brain) contained smaller amounts of DNA at 20 days of gestation. However, the total brain RNA was not reduced. They conclude that maternal protein deficiency started on the day of pregnancy can also result in a decreased number of cells in the fetal tissues.

It is interesting to note that caloric restriction (adequate protein levels) from Days 10 to 20 results in decreased body weight, placental and cerebral weight, DNA, and protein in the offspring. However, if growth hormone is given concomitantly, then the growth retardation is overcome (Zamenhof *et al.*, 1974b). The effect is explained as due to the action of the growth hormone in mobilizing maternal glycogen and fat, the two main sources of energy for the fetus. Growth hormone in normal physiological concentration does not cross the placental barrier (Gitlin *et al.*, 1965) but it is possible that when the levels of this hormone are elevated, some of it passes to the fetus. Alternatively the growth hormone helps to mobilize maternal stores of nutrients and pass them on to the fetus. Whether the growth retardation due to restricted proteins or essential amino acids is reversible by any means remains to be seen. A hypothesis gathering increasing support is that cellular deficiencies, resulting from a low-protein diet during pregnancy, are not restored by an adequate milk supply during the postnatal development. For example, Zeman (1970) conducted elaborate studies in which newborns from mothers on a low-protein diet were suckled by mothers who were fed a control diet. As an added factor the number of pups with each mother was at first eight but after 7 days in one group the number was reduced to four. This was designed to provide an excess milk supply (this may not always be correct, as discussed before in connection with the method of inducing overfeeding). The results indicated that when eight pups were suckling per mother, the decreased organ weights, DNA, and total protein persisted in the postnatal period but did not become severe. Even when the number of pups was reduced to four, there was no change in the cell number but liver, kidney, and heart increased in weight. In the brain, this increased nutritional supply did not result in any significant difference in weight, DNA, or apparent cell size. Thus, whatever deficit was evident in the brain at birth due to protein deficiency during pregnancy, it was not restored even when a normal or extra milk supply was available per pup. This indicates a serious, often debated, implication that there is no provision for "catching up" later, once the cellular deficits are produced.

6.4. EFFECT OF FAT DEFICIENCY

The knowledge about lack of fat in the diet and its effects on health has been based on an extrapolation of animal studies which have been going on for over 50 years. However, in recent years many clinical reports have appeared that prove that there is a close relationship between animal studies and conditions encountered in human populations. Since fats can be synthesized from excess carbohydrates by mammals, lack of fat as such may not produce any abnormal changes in the biochemical or physiological parameter in the short run. But this is not to say that there may be no consequences in the long run. It is well known, from the classic work by Burr and Burr (1930), that EFA cannot be synthesized from other dietary precursors and hence must be provided by the diet. Thus, most of the work related to the role of fat in the development of the CNS is centered around the effects of EFA deficiency on the developing brain.

6.4.1. Nature of the Brain Lipids

The chemical composition of the various organs is believed to be unique in some way and although the exact direct connection between biochemical composition and function of the organ is not completely understood, some progress has been achieved in this direction. It is easier to correlate these two parameters in the case of an organ whose function is well defined, e.g., heart, lung. Brain function is, for the most part, a mystery and, therefore, it is still not possible to assign roles to either its gross or fine chemical composition.

Examination of the basic chemical composition of the brain clearly shows that it is rich both in lipids and proteins but contains very small amounts of carbohydrate (glycogen). In fact, apart from adipose tissue, the normal concentration of lipids in the brain is one of the highest in the body (100 mg/g wet wt.). In terms of dry weight, a little over half of the total weight is made up of lipids. Further, the brain has a unique structural component, the myelin sheath, which on a dry weight basis is 80% lipid and 20% protein. The compositional differences do not stop at the gross constituents but the nature of the brain lipids is also very much different from that found in other organs.

Tables 3 and 4 list the differences in the lipid composition of the liver and the brain.

It is very interesting to note that not only the total lipid content of the brain increases rapidly during the developing period (and the myelin lipids increase during the period of myelination), but some constituents actually decrease or disappear during this period. This orderly change seems to be critical for proper functioning of the CNS. For example, the nondisappearance of esterified cholesterol usually indicates some abnormality with the process of myelination. Thus, only in certain demyelinating diseases does one find an appreciable amount of esterified cholesterol in the brain. Changes in brain lipid composition are brought about not only by lack of EFA in the diet but by lack of other dietary constituents such as proteins as well. At the same time, lack of EFA in the diet can bring about an alteration in the nonlipid components of the brain.

Normally, EFA include linoleic acid ($18 : 2\omega6$) and linolenic acid

Table 3. Differences in the Lipid Composition of Liver and Brain

		Brain lipids			Liver lipids
		Newborn, 27–30	1–15 days, 55–60	Adult, 100–110	
1.	Total lipids (mg/g)				Very small variation with age, 40–50 mg/g
2.	Triglycerides	Found only in trace amounts			Major component ~30% TL[a]
3.	a. Free cholesterol	Adult (20–22 mg/g)			3–4 mg/g ~10% TL
	b. Esterified cholesterol	Small amounts up to 3 weeks			Small amounts
	c. Desmosterol	Only trace adult			Absent
4.	Total phospholipids (mg/g)	12 days, 28		Adult, 47	25–30 mg/g, no age difference
5.	Phosphatidylcholine	Major component ~48–50 mole%			Major component ~ 50% TL
6.	Phosphatidylserine	~11–12 mole%			Minor component ~1% TL
7.	Phosphatidylinositol	Contains mono-, di-, and tri- phosphate inosities			Mostly mono-phosphate irosities
8.	Phosphatidylethanolamine	Diacyl type major component before myelination; alkeneyl-acyl type major after myelination (70%)			Mostly diacyl type
9.	Sphingolipids	Sphingomyelin, cerebrosides, sulfatides			Mostly sphingomyelin
		Very rapid increase during myelination (myelin lipids)			

[a] Total lipid.

TABLE 4. Differences in the Fatty Acid Composition of Liver and Brain Lipids

	Brain lipids	Liver lipids
1.	Hydroxy fatty acids	Minor constituent
2.	Odd number carbon chain fatty acids	Minor constituent
3.	Very-long-chain fatty acids (24 and above)	Minor constituent
4.	Polyunsaturated fatty acids	
	a. Very small amounts of 18 : 2ω6 or 18:3ω3[a]	Sizable amounts of 18 : 2ω6
	b. High amounts of longer-chain polyunsaturated fatty acids 20 : 4ω6 and 22 : 6ω3	Sizable amount of 20 : 4ω6 and lesser amount of 22 : 6ω3

[a] Fatty acids are written as, e.g., 20 : 4, which denotes the total number of carbon atoms followed by the number of double bonds. The ω denotes the position of the distal double bond from the methyl end.

(18 : 3ω3) mainly because the mammalian system was found to be incapable of synthesizing them from commonly available precursors; when excluded from the diet, characteristic symptoms of EFA deficiency were produced. One of the richest sources of linoleic acid is safflower oil (~74% of total fatty acids) whereas soybean and corn oils contain about 50–60% linoleate. Although linseed oil contains as high as 50–55% linolenate, it is generally not used for edible purposes. Soybean oil contains approximately 5% linolenate (Agriculture Handbook, 1979). One of the earliest symptoms of EFA deficiency was established as a characteristic skin condition, growth retardation, and increase in water consumption (Basnayake and Sinclair, 1956) based on the bioassay techniques of Deuel (1951) and Thomasson (1953). Thomasson (1961) also showed an increased permeability of water through an isolated piece of skin *in vitro*. Further work identified other effects like impaired reproduction, erythrocyte fragility, mitochondrial swelling, and increased metabolic rate (Mead and Fulco, 1976).

The question of whether to include both linoleic and linolenic acids as EFA is debatable. The relative effectiveness of linolenate in curing EFA deficiency symptoms is only about one-tenth when compared to linoleate (Mead and Fulco, 1976). Linolenate has to be converted to eicosapentaenoic acid (20 : 5ω3) to be an effective precursor of prostaglandins of the $PGF_3\alpha$ series. This conversion is not a major process and so linolenate is sometimes not considered to be an EFA, at least not in the same way as linoleate and arachidonate. However, the CNS contains an unusually high amount of ω3 (mostly 22 : 6) fatty acids and it is possible that the fatty acids of the ω3 series may be important to the brain. The other specialized tissues that contain high amounts of the ω3 fatty acids are human retina, testis, and sperm (Tinoco *et al.*, 1979).

Most of the early work on effects of EFA deficiency on the developing brain was started by using total fat-free diet. Female rats, usually 1 or 2 weeks before mating, were started on a fat-free diet, and their progeny were examined at various intervals (Paoletti and Galli, 1972). The body and brain weights were severely reduced in all cases. The fatty acid analysis of the brain phospholipids indicated that ω6 fatty acids were decreased and 5,8,11-eicosatrienoic acid (20 : 3ω9) appeared. The formation of 20 : 3ω9 (from 18 : 1ω9) and its accumulation is considered an index of EFA deficiency (Holman, 1973). The fatty acid patterns of myelin lipids, in particular ethanolamine phosphoglyceride (EPG), were also affected indicating that the structure of the myelin may have been altered. Rats given behavioral and performance tests showed that these parameters were also adversely affected by EFA deficiency.

Svennerholm and co-workers (1972) started their dietary regimen much earlier and examined the progeny of both sexes. The diets employed contained the same percentage of fat but low EFA (0.74 cal %), and low (8 cal %), and high (16 cal %) protein. For comparison, controls were fed high EFA (3 cal %) with the low and moderate protein diets. They found that body weight was slightly lower in the low EFA–high protein group, but there were no significant differences in other groups. The fatty acid pattern of the brain was also remarkably similar in all groups but low EFA produced a decrease in cerebroside content (myelin lipid). The 20 : 3ω9 acid increased in the low-EFA group but the 22 : 5ω6 acid increased. These results are in variance with others in this field of research. Perhaps the experimental design by Svennerholm, i.e., adaptation to low levels of EFA for two generations, may explain the apparent paradox.

EFA deficiency is not easy to produce because of the size of the maternal stores. To deplete these stores, a long duration of dietary treatment is needed. Many years back, Dhopeshwarkar and Mead (1961), while evaluating the nutritional effects of *cis,trans*-octadecenoic acid, came to the conclusion that if the diet contains an overabundance of oleic acid, it leads to EFA deficiency even in the presence of small amounts of linoleate. The reason for this is that oleate competes with the utilization of linoleate; for example, under normal conditions, linoleate is converted to arachidonate as shown in Figure 14. [*Note:* The earlier idea of the alternate pathway depicted by broken lines was found to be absent both in the liver (Sprecher, 1975) and in the brain (Dhopeshwarkar and Sub-

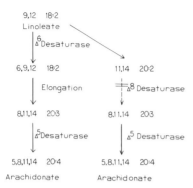

Figure 14. Metabolic pathway of linoleic acid.

ramanian, 1976) when an overabundance of oleic acid is included in the diet and Δ^5- and Δ^6-desaturase systems prefer oleate as substrate and the following reactions predominate:

$$9\text{-}18:1\xrightarrow{\Delta6\text{-desaturase}}6,9\text{-}18:2\xrightarrow{\text{elongation}}8,11\text{-}20:2\xrightarrow{\Delta5\text{-desaturase}}5,8,11\text{-}20:3$$

The final product, eicosatrienoic acid (20 : 3ω9), originating from oleate (Fulco and Mead, 1959), accumulates in EFA deficiency and the ratio of triene/tetraene is used as an index of EFA deficiency (Holman, 1960). Thus, diets containing large amounts of oleate (70–90%) but marginal amounts of linoleate (1–3%) allow normal reproduction, but, due to competitive inhibition by oleate, the utilization of linoleate is inhibited thus producing EFA deficiency rather rapidly.]

This approach was used first to produce EFA deficiency in suckling rats (Menon and Dhopeshwarkar, 1981). The conclusions from this study were (1) Competition from oleic acid toward utilization of linoleate produces dietary EFA deficiency; (2) saturated fatty acids do not act similar to oleic acid; (3) EFA deficiency produced this way is more severe in males than females in the progeny; (4) 21 days after birth the triene (20 : 3ω9) was found in appreciable amounts in the liver and serum of the pups but the amount in the brain was relatively low; (5) the progeny from rats on either coconut oil (saturated fat), laboratory chow, or safflower oil did not show any trace of 20 : 3ω9 in their tissue lipids; (6) after 90 days the triene/tetraene ratio was 10–15 times higher in brain total phospholipids as well as in lecithin, in the EPG and choline phosphoglyceride components the ratio was 2–4 times higher when rats of the oleic acid group were compared with those on coconut oil; (7) the brain subcellular fractions, microsomes, and mitochondria also showed an accumulation of 20 : 3ω9, the same was true in the case of myelin, particularly the EPG component; (8) the lipid/protein ratio was found to be reduced by about 50% in the brain microsomes from rats on the high-oleic acid diet. Finally, the myelin EPG odd-chain fatty acids showed a dramatic decrease indicating that the myelin structure may have been altered in EFA deficiency. Here, one can see that EFA deficiency produced by an imbalanced fatty acid diet altered many biochemical parameters of brain physiology and structure. The competitive role of oleic acid in the utilization of linoleate was further proved in a study by Menon *et al.* (1981b). It was found that if the oleic acid content in the dietary fat was increased there was a concomitant need for higher amounts of linoleate. For example, a minimum of 1% of the total calories by way of linoleate is considered adequate for normal growth and development (Holman, 1960). It was found that when the diet contained mostly saturated fat and only 0.1% of the total calories came from linoleate, even after 90 days, the accumulation of the 20 : 3ω9 (EFA deficiency index) in the brain was negligible. However, when oleate provided most of the total calories (high-oleic-acid diet) even the presence of 1% of the calories as linoleate was not enough to prevent accumulation of 20 : 3ω9 in the brain. Thus, it was concluded that there is an increased requirement of linoleate in the presence of dietary abundance of oleate because of the competitive nature of oleate. No such competition is caused by saturated fatty acids.

This method of inducing EFA deficiency was further exploited by Menon *et al.* (1981a) to study EFA deficiency in fetuses. Rats were fed a high oleic–low linoleic acid diet for 4 months and mated. On Day 21 of gestation, fetuses were removed along with placental tissue and maternal blood and liver tissue. The fetuses, when compared to controls, were smaller and contained higher amounts of 18 : 1 and 20 : 3 indicating an EFA deficiency state. The weights of fetuses born at term (21 days) to dams on a high oleic–low linoleic acid diet were equal to those of 18-day fetuses from control dams; in other words, there was a growth retardation of 3 days, which for a species that has only a short 21-day gestation period is a serious deficit. The small-for-gestational age newborn has many handicaps to overcome for successful survival and normal growth as mentioned in connection with FAS. The animal model could be used to study the human small-for-date babies. Decreased amounts of fatty acids of the ω6 family in fetal and maternal tissues were observed; however, the decrease was relatively small in placental tissue. This suggested that placental tissue does not seem to be affected to the same extent as the fetus. It also indicates that placental tissue has the capacity to extract and maintain a higher level of ω6 fatty acids. This may be viewed as a possible mechanism of preserving the ω6 fatty acids for prostaglandin synthesis needed for inducing labor and other vascular changes in the fetus.

The dietary treatment of female rats before mating, during gestation, and in the lactating period is important to produce EFA deficiency in the offspring. The newborn depends on the milk supply for all its nutrition and for the next 10–15 days the growth of the pups depends on an adequate milk supply. Lack of fat in the diet or low levels of linoleic acid in the diet does not seem to actually reduce the quantity of milk during lactation but the linoleic acid content of the milk is considerably reduced by the deficient diet (Galli, 1972; Menon *et al.*, 1981b). Thus, the growing animal continues to receive very low amounts of linoleate during the important period of lactation in which myelin synthesis and deposition is proceeding at a rapid rate.

Crawford and Sinclair (1972) have gone one step further, i.e., fed rats from weaning on a diet containing an extremely low fat content with EFA providing 0.33% of the calories (ratio of 18 : 2 to 18 : 3 was 5 : 1). Under these conditions the mortality in pups born to these dams was very high and only 10% of the pups survived. Thus, selectively, only the very hardy are available for study. The newborns have smaller brains but the same concentration of lipid per gram of fresh weight. A very interesting observation that the lack of fat or EFA brings about a different pattern of tissue fatty acids in guinea pigs was made by these authors. They point out that the ratio of 22 : 5ω3 to 22 : 6ω3 was reversed in guinea pigs which is quite different from what one sees in the rat. Thus, it is important to note the species difference.

There is uncertainty and debate on the comparative effects of protein deficiency and EFA deficiency. The former is considered to produce much more permanent damage, not reversible by the intake of an adequate diet during adulthood, whereas the effects of EFA deficiency are apparently reversible. However, this conclusion was reached from observations linked to restoration of lipid profiles in the whole brain and subcellular fractions in rats raised on an EFA-deficient diet up to a month or more after birth followed by a balanced adequate diet during later adult life. No one, of course, knows whether the restoration of the lipid profile actually means "repair of damage." The restoration of lipid profiles was studied by White *et al.* (1971) who reported that rats raised on essentially a fat-free diet for the first 90 days followed by an adequate diet for an additional 90 days were able to restore lipid profiles completely. Upon substitution of an adequate diet, the ω6 family fatty acids rebounded dramatically and the excess ω9 family fatty acids that had accumulated during the deficient diet disappeared after diet restoration.

Essentially similar results were obtained by Sun *et al.* (1975) in mice. These authors report that in developing mice the half-lives of ω9 polyunsaturated fatty acids of total EPG of brain microsomal, synaptosomal, and myelin fractions were 3, 10, and 15 days, respectively. Further, in general, the disappearance of 20 : 3ω9 fatty acid was faster in diacyl EPG as compared with ethanolamine plasmalogens. Thus, the restoration of the brain lipid profile seems to be possible, but recovery from deleterious effects that may have been caused during active growth and development may still be present. Our limited ability to accurately measure these effects, in test animals, seems to be inadequate to pass any final judgment.

The unusually high amount of fatty acids of the ω3 family in the CNS was pointed out earlier and this fact has puzzled many workers in the field of nutrition. Even though the amount of ω3 fatty acids (mostly 18 : 3ω3) in the diet is usually very small or many times totally absent, it takes a very long time to deplete this from the brain. Further, the half-life of these highly unsaturated fatty acids is rather short, on the order of 15 days (Dhopeshwarkar and Mead, 1975). It is, of course, possible that in certain compartments (similar to cholesterol) the half-life may be much longer. But such compartmentalization in the case of docosahexaenoic acid, for example, has not been clearly established. Thus, if the dietary supply of the precursor 18 : 3 or product ω3 family fatty acids is cut off, how does the brain maintain this high level? One possible explanation lies in the fact that there is a preferential uptake of 18 : 3ω3 fatty acids by the developing brain. Experiments with ^{14}C-labeled 18 : 3 and 18 : 2 acids by Dhopeshwarkar and Mead (1973) showed that there is a preferential uptake of linolenic acid (18 : 3ω3) over linoleic acid (18 : 2ω6). The importance of the linolenic acid family fatty acids to the CNS has received special attention by Bernsohn and Stephanides (1967) who originally proposed a possible connection between dietary deficiency of EFA and multiple sclerosis. Later studies by Bernsohn and Spitz (1974) have hinted that linolenic acid may have a biochemical function distinct from that of linoleic acid in restoring certain membranebound enzymes.

Since linoleic acid (ω6 family fatty acid) suppresses the conversion of 18 : 1 to 20 : 3 (ω9 family) attempts have been made to see if high levels of 18 : 3 (ω3) in the diet would lead to decreased levels of 20 : 4 (ω6). In fact, just as the 20 : 3ω9/20 : 4ω6 ratio is an index of linoleic acid deficiency, Galli *et al.* (1974) have proposed the 22 : 5ω6/22 : 6ω3 ratio as an index of linolenic acid deficiency. Taking both of these ratios into consideration, Galli *et al.* (1976) suggest keeping the linoleate:linolenate ratio in the diet very similar to that found in milk lipids of lactating females fed a normal adequate diet. The ratio is 12 and 6 in rabbit and guinea pig milk, respectively, but around 3 in human milk (Smith *et al.*, 1968; Crawford *et al.*, 1973).

A strict dietary requirement of linolenic acid (ω3) has been shown only in the case of rainbow trout (Sinnhuber *et al.*, 1972) and was established as 1% of dry weight of the diet. Lack of 18 : 3 in the diet of trout caused poor growth, elevated tissue levels of ω9 fatty acids (including 20 : 3ω9), necrosis of the caudal fin, fatty and pale liver, dermal pigmentation, increased muscle water content, increased mitochondrial swelling, heart myopathy, and lowered Hb level. Addition of ω6 fatty acids did alleviate some of the above symptoms but not all of them. In fact linoleate seems to aggravate some symptoms such as heart myopathy. No such absolute requirement of the ω3 fatty acids has ever been shown in rats or other mammals. In fact, even after depriving ω3 fatty acids for three generations neither fertility, organ weight, or survival rate was affected in any significant way (Tinoco *et al.*, 1971).

Recently, eicosapentaenoic acid (5,8,11,14,17–20 : 5), which is an intermediate in the conversion of 18 : 3ω3 to 22 : 6ω3, was shown to be converted to prostaglandin PGE$_3$ and thromboxane A$_3$ (Needleman *et al.*, 1979). The triene prostaglandins have a different functional activity from that of the prostaglandins formed from ω6 fatty acids. Thus, ω3 fatty acids may have a totally different biological function including that related to brain physiology.

Lack of ω3 fatty acids in the diet may affect another specific tissue that contains very large amounts of ω3 fatty acids. This tissue is the retina where 22 : 6ω3 occurs in large amounts, about 35–55% in membranes, that make up the photoreceptor outer segments (Stone *et al.*, 1979). In the retinal EPG one finds 75-100% of the fatty acids in the 2 position are ω3 fatty acids (Anderson and Sperling, 1971). If rats are fed fat-free diets from weaning (21 days), only about 10–12% of 22 : 6ω3 acid is reduced in the retina lipids (Futterman *et al.*, 1971). However, if one uses 1.25% of highly purified 18 : 2 acid as the only source of fat, depletion of ω3 fatty acids was greater. Tinoco *et al.* (1977) using this approach showed that in the first generation there was a decrease of 60% and almost 85–90% in the second generation. With such dramatic decreases in ω3 fatty acids in the retina, one would expect some abnormalities in the physiology of vision. Wheeler *et al.* (1975) did find a reduction in the electrophysiological response in the retina of ω3 fatty acid-depleted rats. Further, the inclusion of ω3 fatty acids in the diet brought a rapid reversal of the decreased response. The authors point out that biological membranes can be stable and relatively rigid like the myelin and these are generally rich in saturated and monounsaturated fatty acids; on the other hand, active membranes like those of mitochondria, synapse, and photoreceptor outer segments contain relatively higher amounts of ω3 PUFA. Thus, lack of ω3 fatty acids may affect these cellular membranes much more than other membranes (Fleisher and Rouser, 1965).

No disease condition in humans has been reported due to the lack of ω3 fatty acids in the diet. However, Clausen and Möller (1967) have proposed a hypothesis regarding the etiology of multiple sclerosis; it is based on another theory that a measleslike virus is the causative agent of multiple sclerosis. Since all those who contract measles do not develop multiple sclerosis, there might be another factor needed to produce multiple sclerosis. This factor may well be nutritional in nature. The autoimmune disease allergic encephalitis (considered as an animal model to study multiple sclerosis by some workers) can be produced in animals that are deficient in ω3 fatty acids with a much smaller dose of the antigen. Thus, a combination of exposure to the virus and concurrent nutritional deficiency are hypothesized as factors in the etiology of this disease.

Yet another hypothesis based on a nutritional factor, arachidonic acid 20 : 4ω6 in particular, has been proposed by Mickel (1975) for the etiology of multiple sclerosis. The hypothesis is based on lipid peroxidation, a normal process occurring in most cell membranes and inhibited by antioxidants such as α-tocopherol. The lipid peroxides are a highly reactive species and can cause protein denaturation, e.g., by attacking the sulfhydryl groups of the enzymes. The hypothesis assumes that such lipid peroxides arise in the G.I. tract. Normally, these are not absorbed very readily; however, under specific conditions such as intestinal infection, viral or otherwise, the peroxides could be absorbed much more readily. Another contributing factor may be that the infection may release lysosomal peroxidase that may increase the peroxidation of lipids. In either case a diet rich in PUFA (20 : 4ω6) but low in vitamin E would be much more damaging. The peroxidized arachidonic acid via prostaglandins and related compounds can cause platelet aggregation and, in fact, such compounds are found associated with the platelet surfaces. With adherence of platelets to postcapillary venules, the peroxidized lipids might be released and pass across the capillary endothelial cells and attack adjacent oligodendroglia. The plaques of multiple sclerosis could be the result of protein denaturation. The denatured proteins might serve as an antigenic stimulus in the development of an autoimmune process which is believed to be the underlying process leading to multiple sclerosis. This hypothesis needs careful step-by-step research to reach final conclusions.

6.5. EFFECT OF TRANS FATTY ACIDS

Edible vegetable oils used in our daily cooking contain triglycerides that have unsaturated fatty acids with *cis* configuration. However, for many years these vegetable oils have been commercially purified, bleached, and hydrogenated. The first two processes remove some of the unwanted and harmful impurities, e.g., gossypol found in trace amounts in cottonseed oil (Dhopeshwarkar, 1981). However, the last process, partial hydrogenation, transforms the spatial configuration of double bonds from *cis* isomers to *trans* isomers. The *trans* isomers of EFA lose their "essential" nature; in other words, the fatty acids with double bonds in the same position as EFA (9,12 in the case of linoleate) are changed from the somewhat bent configuration of *cis* forms to the straighter spatial configuration of *trans* bonds and can neither cure symptoms of EFA deficiency nor provide precursors for biologically potent compounds such as the prostaglandins. Thus, because of commercial hydrogenation, one can expect increased shelf life and the possibility of making multiple products; but the process essentially destroys the "essential" nature of vegetable oils. Newer methods to overcome this effect include "controlled" conditions to reduce formation of *trans* bonds and blending with unhydrogenated oil to increase the PUFA content. The complex mixture of fatty acids (due to "wandering" of the olefinic centers) found in partially hydrogenated products can be guessed from the following example: linoleic acid (9-*c*,12-*c* 18 : 2) can give rise to 9-*t*,12-*c*; 9-*t*,12-*t*; and 9-*c*,12-*t* 18 : 2; additionally it is known to provide positional isomers in which the double bond migrates between the methyl and carboxyl carbons (Dutton, 1979).

The nutritional effects of a *trans* fatty acid diet are controversial and have been discussed by Alfin-Slater and Aftergood (1979); opinions range from "no harmful effects" to "implication in incidence of cancer" (Enig *et al.*, 1978). Even the claim that a *trans* fatty acid diet causes elevation of blood cholesterol (Vergroesen and Gottenbos, 1975) has been disputed (Heckers *et al.*, 1977). That *trans* fatty acids seem to readily cross the placental barrier and get deposited in the fetal tissue was proved by using radioactive *trans* fatty acids (Moore and Dhopeshwarkar, 1980). If female rats are fed partially hydrogenated fat 2 weeks before mating and continued on this diet during pregnancy and lactation, the offspring are born with *trans* and positional fatty acid isomers in most of their tissues, including the brain. Since the isomers are secreted in the milk, the suckling pups soon build up the total *trans* fatty acid content (Menon and Dhopeshwarkar, 1981). Further, it has been shown that the radioactive *t,t* 18 : 2 administered to pregnant rats ends up in the 2 position of fetal brain lecithin (Moore and Dhopeshwarkar, 1981). Similar results were obtained with 12-day-old rats when the isotope was injected intracranially (Karney and Dhopeshwarkar, 1979). The importance of this finding lies in the fact that lecithin (and other phospholipids) is a component of all cell membranes and that fatty acids of different phospholipids are selectively esterified either at the 1 or 2 position. Under normal physiological conditions, in most (but not all) tissues PUFA are found in the 2 position. These are, again, under normal conditions, *cis* in configuration. But when a diet contains *trans* fatty acids, these displace part of the normal *cis* acids. This might be considered to have two effects. One is that, since the physicochemical properties of *trans* fatty acids are different from those of their *cis* counterparts, the membrane character may be altered. The second effect may be due to the fact that phospholipids are substrates for lypolytic enzymes such as phospholipase A_2 which specifically hydrolyze and set free the fatty acid at the 2 position. If this happens to be linoleic or arachidonic acid, then it becomes a precursor for prostaglandin synthesis. If a *trans* fatty acid occupies this position (since these are not precursors for prostaglandin synthesis), a decrease in prostaglandin levels can be predicted in animals receiving a diet containing *trans* fatty acids (Kinsella, 1979).

Although such physiological effects are possible, as yet, no deleterious effects can be specifically connected, in human populations even after 30–40 years of dietary intake, to any specific pathological condition. In these circumstances, more attention should be directed toward the indirect effects of enhancing EFA deficiency by dietary *trans* fatty acids (Hill *et al.*, 1978). Recently Schrijver and Privett (1981) have reported that prolonged feeding (12 weeks) of *trans* fatty acids depressed the Δ^6-desaturase activity by 20% and elevated the Δ^9-desaturase activity threefold in liver microsomes. Our own studies (Karney and Dhopeshwarkar, 1978) showed that although *t,t* 18:2 was desaturated to 6-*c*,9-*t*,12-*t* 18:3 and elongated to 8-*c*,9-*t*,12-*t* 20:3, the last step, converting it to a tetraene isomer, was blocked, i.e., the 8-*c*,9-*t*,12-*t* was not a suitable substrate for the Δ^5-desaturase. Therefore, the tetraene isomer was not formed. Thus the interference of *trans* fatty acids in the metabolism of linoleic acid (EFA) seems to be established and may exert an influence on normal utilization of EFA. In the case of borderline EFA deficiency this suppression by *trans* fatty acid may aggravate EFA deficiency and its consequences on brain metabolism.

One of the most intriguing components of brain lipids, collectively termed gangliosides, also have a very complex structure. Gangliosides contain fatty acids (FA), sphingosine (SPH), hexoses (galactose, Gal; glucose, Glu), an *N*-acetylhexose (*N*-acetylgalactose, GalNAc), and sialic

acid (*N*-acetylneuraminic acid, NANA). However, the number and structural attachment can vary. For example, one of the complicated gangliosides G_{alb} has the following structure:

$$
\begin{array}{l}
\text{Gal}(1 \to 3)\text{GalNAc}(1 \to 4)\text{Gal}(1 \to 4)\text{Glu}(1 \to 1) - \text{SPH} \\
\quad 3 \qquad\qquad\qquad\qquad\quad 3 \qquad\qquad\qquad\quad | \\
\quad \uparrow \qquad\qquad\qquad\qquad\quad \uparrow \qquad\qquad\qquad\quad \text{FA} \\
\quad 2 \qquad\qquad\qquad\qquad\quad 2 \\
\text{NANA}(8 \leftarrow 2)\text{NANA} \quad \text{NANA}(8 \leftarrow 2)\text{NANA}
\end{array}
$$

The nomenclature of gangliosides is also confusing since various groups
working in this area have used different abbreviations. The notation
used by Svennerholm seems to be most popular (Svennerholm, 1976).
The gangliosides have been found in all the cells, subcellular particles,
and fluid compartments of the brain, although some cells contain more
than others. For example, the processes and perikarya of the neuron
contain a large bulk of gangliosides in the brain (Ledeen *et al.*, 1976).
Another possible function attributed to the ganglioside (G_{m1}) is that it
acts as a natural receptor for cholera toxin (Svennerholm, 1976). Some,
as yet unclear, relationship of gangliosides to synapses has been suggested. In short, the exact function of gangliosides in the brain remains
unknown. Merat and Dickerson (1974) showed that rats on a low-protein
diet (7% protein diet) contained reduced amounts of gangliosides in
each of the major brain areas, such as the forebrain, brain stem, and
cerebellum. These authors imply that the reduction in gangliosides indicated decreased dendritic arborization. Such an effect as shown in
histological studies by Cragg (1972). Yusuf and Dickerson (1978) studied
the effects of malnutrition (50% decrease in food intake compared to
controls) on gangliosides of the forebrain, brain stem, and cerebellum
of Wistar rats. Their findings indicate that undernutrition during gestation and up to 21 days after birth did not produce any significant
change in brain gangliosides. However, when continued up to 121 days,
it resulted in a significant decrease of lipids in each of these brain regions.
For example, ganglioside G_{dlb} in brain stem and G_{tla} in cerebellum all
decreased when compared to controls. During the first 21 days, a slight
effect of undernutrition was observed in the cerebellum but not in other
areas, indicating that there might be a greater vulnerability of the cerebellum to undernutrition during the early period of life. The results,
showing an effect of malnourishment during the later period of life, 21
to 120 days, are different from those commonly found by others in this
field. It seems that the brain, even after the growth spurt is over, is

vulnerable to nutritional stresses at least with respect to ganglioside concentrations.

So far we have discussed the effects of EFA deficiency on the lipids of the brain, but recently McKenna and Campagnoni (1979) studied the effects of EFA deficiency on brain DNA, RNA, and protein content. These authors found a decrease in all of the above and suggested that there was a retardation of brain growth and cell number of about 1 week compared to controls. The DNA content of both the controls and EFA-deficient animals became comparable at 20–22 days, probably due to the outgrowth of glial cells, but brain protein and RNA content remained lower in deficient mice at all ages. The myelin lipid constituents were also drastically decreased. Proteolipid protein was significantly reduced. The effect of a low-protein diet on lipid parameters has not received much attention from researchers in this field.

6.5.1. Erucic Acid

Another very controversial dietary fat has recently received a great deal of attention. The oil from which it is derived is rapeseed oil, from a plant belonging to the order Rhoedales of the family Cruciferae. In tropical countries like India, it is called mustard oil. These oils have been part of the human diet in Canada, Argentina, India, Poland, Germany, and Sweden. Canadians consumed more rapeseed oil than any other edible oil during the 1972–1974 period (Singer, 1977). Rapeseed oil contains a fatty acid, erucic acid, which has 22 carbons and a double bond and belongs to the $\omega 9$ family. Before attempts were made by plant geneticists to reduce the concentration of this fatty acid (Slinger, 1977), the concentration of erucic acid was approximately 40–60% and this is now called the high-erucic-acid rapeseed oil. This oil, when fed to rats, produced myocarditis and hepatopathological lesions. The causative agent was believed to be erucic acid, because feeding trierucin also produces similar lesions (Beare-Rogers et al., 1972). An isomer of erucic acid known as cetoleic acid (22 : 1, $\omega 11$) found in marine oils was no different. Recently, a selective isolation of mutant strains yielding an oil with essentially no erucic acid has provided commercial quantities of low-erucic-acid rapeseed oil. One such Brassica napus variety "tower" is now licensed for sale in Canada. The decrease in erucic acid was compensated by an increase in oleic, linoleic, and linolenic acids. Now this variety has approximately 95% total fatty acids with an 18-carbon chain length. Rocquelin and Cluzan (1968) found that feeding of low-erucic-acid rapeseed oils to rats did not prevent cardiac lesions. Kramer et al. (1975) found that the cardiopathogenicity was associated with the triglycerides and not the other components of the oil such as brassicasterol or glucosinolates that are present in refined low-erucic-acid rapeseed oil. The increase in linolenic acid in low-erucic-acid oil may be a factor in pathogenicity (McCutcheon et al., 1976). The low-erucic-acid oil gave only a slightly higher severity rating than corn oil (Slinger, 1977). Another method of reducing cytotoxicity of this oil was to subject it to hydrogenation. Although this eliminated linoleate and linolenate from the oil, it introduced positional and trans isomers of erucic acid (considerably reduced compared to original) and oleic acids. Beare-Rogers (1979) points out that cardiac pathogenicity of low-erucic-acid rapeseed oil is further reduced by using partially hydrogenated oil, a feature clearly desirable in human nutrition. But one should be aware that the EFA content found in these low-erucic-acid rapeseed oils is lost in partially hydrogenated oils; therefore, there would be an increased need for linoleic acid. Thus, blending with soybean or corn oil would be desirable.

The amount of erucic acid found after feeding rapeseed oil to rats is very low in the brain when compared to other organs; therefore, most of the studies have been concentrated on organs other than the brain. In fact, the question of why erucic acid should prove to be toxic when its elongated product, nervonic acid (24 : 1ω9) of the same family of fatty acids (a naturally occurring constituent of brain lipids), is perfectly compatible with physiological functions of the CNS remains unanswered (Dhopeshwarkar, 1981). In this connection, the experiments by Lall and Slinger (1973) seem to indicate that rapeseed oil with high-erucic-acid content fed to laying hens exerted a depressing effect on egg production, egg weight, yolk weight, and egg hatchability. No similar studies have been reported on the growth and development of the CNS in rats during intrauterine or suckling periods.

6.6. ROLE OF CARBOHYDRATES IN BRAIN DEVELOPMENT

The three major macronutrients of our diets are carbohydrates, proteins, and lipids. Although the carbohydrate content of most plant-derived foods is high and occupies a major role in the overall diet of the general population (except in some groups like Eskimos and other strict meat eaters whose diet is low in carbohydrates and high in fat and protein), the role of carbohydrates in health has not received much attention. In fact, the lack of carbohydrates is sometimes equated with starvation. Only in recent years have carbohydrates and, in particular, certain types of carbohydrates become a subject of intense research. Thus, diets high or low in various types of carbohydrates are increasingly included in experiments just like other components such as proteins and lipids. People who gain weight rather quickly tend to avoid carbohydrates.

Dietary carbohydrate sources include cereals, sugars, starchy fruits (e.g., dates, figs, bananas) roots, tubers, and a variety of products made from these. Cereal alone contributes about 40% of total human food intake, although varying in different areas of the world. Sugar consumption has also increased during the last three decades. An average meal provides 60% starch, 10% lactose, and 30% sucrose, all of which finally produce monosaccharides, glucose being the most prominent, followed by fructose and galactose. On the average, even though different amounts of carbohydrates are included in the diet providing varying amounts of glucose, the blood glucose level does not increase or decrease proportionately. Blood glucose levels may start at 80 mg/dl and rise to about 120 mg/dl approximately 60 min after an average meal; there is then a prompt decrease and a return to base level within about 3 hr.

The hormone secreted by the β cells of the islets of Langerhans of the pancreas, which regulates blood sugar, is insulin. Insulin deficiency is a common and serious pathogenic condition in human populations. Diabetes mellitus caused by insulin deficiency is characterized by polyuria, polydipsia, weight loss in spite of increased appetite, hyperglycemia, ketosis, acidosis, coma, and ultimately death. Usually the biochemical lesion can be associated with decreased entry of glucose into peripheral tissue cells and increased hepatic gluconeogenesis followed by increased secretion into the circulation. The action of insulin related to increased uptake of glucose can be seen in the skeletal muscle, adipose tissue, and liver, but, surprisingly, no such effect can be seen in the kidney tubules or the brain (except some parts of hypothalamus). Thus, glucose uptake in the brain is normal in a diabetic individual. A diabetic person who is excreting glucose in the urine loses 4.1 cal for each gram of glucose excreted. To cover this energy loss, the food intake goes up (increased appetite) and this raises blood glucose even more, leading to further urinary loss. Since the cells no longer get the supply of glucose for energy, they utilize fats and protein. Thus, carbohydrate nutrition plays a key role in nutrition of both proteins and lipid; all three macrocomponents are physiologically interrelated.

6.6.1. Changes in Lipid Metabolism

The principal abnormality caused by lack of insulin is increased catabolism of lipids and decreased synthesis of fatty acids and triglycerides. The increased catabolism of lipids leads to increased amounts of ketone bodies. Under normal conditions (normal insulin level and response to dietary carbohydrate), 50% of ingested glucose is burned to CO_2 and H_2O, 5% is converted to glycogen, and the remaining (30–40%) is converted to fat and stored in the depots. In the presence of a decreased level of insulin only 5% or less is converted to fat; at the same time, there is no increased conversion to glycogen and the amount of sugar burned to CO_2 is decreased. Thus, glucose accumulates in the blood.

Insulin inhibits hormone-sensitive lipase in the adipose tissue and thereby regulates the FFA level in blood. In the absence of insulin, the FFA level in the circulating blood plasma increases. The higher amounts of FFA entering the liver can be oxidized to acetyl-CoA, and this leads to further production of ketone bodies.

6.6.2. Changes in Amino Acid Metabolism

A decreased level of insulin accelerates the catabolism of amino acids and increases the conversion of amino acids to glucose. A lack of insulin increases blood amino acid levels and the administration of insulin, or insulin secreted in response to a high carbohydrate meal actually decreases blood amino acid levels. This is particularly true of neutral amino acids (un-ionized), e.g., Leu, Ile, Val, Phe, and Tyr. (This, as we will see later, is a very important reaction of insulin that ultimately leads to a higher level of tryptophan and 5HT in the brain.) Alanine is converted to pyruvic acid leading to gluconeogenesis during periods of decreased insulin secretion. Other important enzymes leading to synthesis of glucose, such as phosphoenolypyruvate carboxylase (oxaloacetate → phosphoenolpyruvate), fructose-1,6-diphosphatase (fructose diphosphate → fructose 6-phosphate), and glucose-6-phosphatase (glucose 6-phosphate → glucose), also increase in activity.

In summary, insulin is responsible for regulating blood glucose levels within a range of about 80–150 mg/dl during pre- and postprandial periods. Insulin deficiency causes decreased entry of glucose in many tissue cells and increased synthesis of glucose in the liver. Glucose may be formed from protein, and energy, therefore, is made available from catabolism of proteins and fats. The blood level of FFA and ketone bodies increases. Only insulin can reverse all of these conditions. Insulin excess can also occur but is not as common as insulin deficiency. Insulin excess will rapidly cause hypoglycemia and since glucose is a preferred fuel for brain function, hypoglycemia in extreme conditions can lead to convulsions, coma, and death. The usual compensatory mechanism is stimulation of epinephrine release causing glycogenolysis and release of glucocorticoids to increase gluconeogenesis in order to combat hypoglycemia. All these reactions have one goal—to quickly provide glucose for the brain.

Theoretically a continuous prolonged hyperglycemia can exhaust the β cells and stop secretion of insulin. The cells can even die but the pancreatic reserve is large. Normally glucose exerts a feedback control on the secretion of insulin by the pancreas. Glucose enters the pancreatic cell quite readily without the need of insulin. Levels above 110 mg/dl of rat blood will begin stimulation to increase insulin secretion. Initially, there is a rapid increase in secretion followed by a slower prolonged increase. The feedback control controls the levels of glucose and insulin in unison. Certain amino acids (ketogenic, like leucine) and ketone bodies stimulate insulin secretion; in turn, insulin promotes amino acid incorporation into proteins.

From the above discussion it is quite clear that carbohydrate metabolism is so interwoven with insulin (also glucagon and other hormones) that one cannot separate them. A dietary deficiency of carbohydrate is unlikely to occur but a dietary surplus of carbohydrates (or sugars) is quite common. Since the level of insulin also correlates with the intake of carbohydrate, we should consider them together. It is well known that the incidence of diabetes due mainly from deregulation of insulin production and/or secretion is rather high and has become a worldwide subject of great interest and anxiety.

Diabetes during pregnancy, when not strictly managed, is known to result in a higher incidence of perinatal mortality including stillbirths and death during the first 10 days of life (Pederson et al., 1974). The first and foremost goal should be to have a baby of normal weight at birth. A small-for-date baby and a premature baby both face an uphill struggle. It has been reported that in a study which included 1452 infants (with a birth weight of <1000 g) the number of malformed infants was 116. The overall frequency of congenital malformations was about three times higher than that of the controls (Pederson, 1975). The improvement that is being achieved in recent years is mainly due to improvement of metabolic control during pregnancy and far more intensified care of the newborn. Although debatable, it has been reported that newborns of diabetic mothers have a higher incidence of hyperinsulinemia compared to controls (Pederson et al., 1974). Researchers reason that functional maturation of β cells is accelerated due to maternal hyperglycemia. Bibergeil et al. (1975) mention that metabolic control during pregnancy, treatment of toxemia, and control of urinary tract infections during pregnancy can lead to the prevention of somatic maldevelopment as well as psychonervous symptom maldevelopment in diabetic offspring. The incidence of diabetes in children born to diabetic mothers is only 0.7% but this is 20 times higher than in controls. In general then, among the survivors, the incidence of metabolic disorder is higher and needs to be treated for the rest of their lives.

6.7. HYPOGLYCEMIA

Infants of diabetic mothers or mothers with toxemia related to pregnancy and those with hereditary metabolic or endocrine disorders are obviously more likely to suffer from hypoglycemia. According to a report by Chase *et al.* (1973), the incidence of hypoglycemia in term infants was about 10% but as high as 25% in small-for-date babies. In prematurely born and SGA babies the incidence is believed to be as high as 67% (Lubchenco and Bard, 1971). Neurological damage has been observed following autopsy of children who died from severe hypoglycemia (Anderson *et al.*, 1967). Several reports have also been published dealing with experimental animals, e.g., rats (Jones and Smith, 1971), where neurological damage has been shown in the cerebral cortex and the hippocampus. Brierley *et al.* (1971) showed similar damage in monkeys. According to the report by Chase *et al.* (1973), if undernutrition is not superimposed, hypoglycemia alone (induced by insulin injection to the pups) did not produce reduction in body weight. However, myelination was affected, as indicated by reduced incorporation of ^{35}S into brain sulfatides. Other lipids such as total phospholipids and cholesterol were also decreased in the brain but this may be a reflection of decreased cellularity. It may be important to point out here that insulin does not cross the blood–brain barrier; therefore, hypoglycemia or decreased glycogen stores may be due to an actual fall in blood glucose. In the case of humans, it may be important to know the duration and number of hypoglycemia episodes and the age (pre- and postnatal) to draw firm conclusions and clarify the type of neurological damage.

6.8. HEXOSES OTHER THAN GLUCOSE

6.8.1. Fructose

Fructose is claimed to be less irritating to the veins than glucose and more readily taken up by tissues (brain not included). However, its renal threshold is lower and so any increase in blood fructose levels will soon spill out in the urine. Fructose is considerably sweeter, 70% more than sucrose; therefore, a smaller amount is needed for sweetening. This makes it suitable for decreased caloric intake and subsequent weight control. At the same time, fructose is readily converted to glucose in the liver. Fructose may be phosphorylated to form fructose 6-phosphate. However, the affinity of the enzyme for fructose is low compared to its affinity for glucose. Therefore, the major reaction is phosphorylation of fructose to fructose 1-phosphate catalyzed by a specific kinase. The activity of this enzyme is not affected by insulin; therefore, it was suggested as a substitute carbohydrate source for diabetics. Fructose is cleared from the blood at a normal rate even by insulin-dependent diabetics. However, appreciable quantities of fructose can be metabolized this way only in the intestine and in the liver; therefore, fructose treatment is of very limited value. Fructose 1-phosphate is split into two triose units, but, unlike fructose 1,6-diphosphate which yields glyceraldehyde 3-phosphate, free glyceraldehyde is formed. Then the triose kinase converts free glyceraldehyde into glyceraldehyde 3-phosphate. The two trioses, dihydroxy acetone phosphate and glyceraldehyde 3-phosphate, can either enter the glycolytic pathway (EM pathway) or may combine under the influence of aldolase and be converted to glucose. Most of the fructose is converted to glucose by this pathway in the liver.

Fructose as such cannot sustain the normal metabolic activity of the brain. One of the reasons, as mentioned before, is that fructose cannot be taken up by the brain rapidly enough, at least in the adult. During infancy, fructose may have a better chance due to a lower metabolic demand or the so-called "incomplete" blood–brain barrier.

Persons who lack the aldolase specific to fructokinase (genetic error) cannot tolerate fructose and excrete a major portion of ingested fructose in the urine. Animal experiments indicate that a drug-induced increase in the concentration of fructose 1,6-diphosphate can lead to convulsions, probably due to an interruption of glycolysis (Sacktor *et al.*, 1966).

6.8.2. Galactose

Metabolically, galactose is phosphorylated to galactose 1-phosphate which is converted to glucose 1-phosphate and then to glucose 6-phosphate and incorporated into the pool generated from glucose. Galactose and glucose are readily interconvertible under normal conditions and behave very similarly. Most of the galactose found in the brain is a component of sphingolipids (cerebrosides, sulfatides, and gangliosides). In galactosemia, a genetic disorder, galactose 1-phosphate uridyltransferase activity is either missing or is very low. In the patients with this disease, jaundice or hepatomegaly occurs early, cataracts may develop soon after birth, and, if untreated, mental retardation occurs. Blood galactose is elevated and excreted in the urine. This may be used as an early diagnostic test. Red blood cell (RBC) galactose 1-phosphate uridyltransferase enzyme activity can also be used as an *in vitro* enzymatic test. In rat experiments, high levels of galactose induced by intravenous injections have failed to mimic the human genetic disorder, but in chickens cerebral dysfunction has been produced (Kozak and Wells, 1971). It is assumed that high levels of galactose reduce glucose uptake in the brain. The chick recovers from diet-induced hypergalactosemia when treatment in this disease is discontinued. In humans, diet does not play any important role.

6.9. POLYSACCHARIDES: GLYCOGEN

Unlike liver and muscle the brain stores of glycogen are small (3.3 μmole/kg rat). For many years the exact concentration of glycogen in the brain was in doubt. The reason was rapid breakdown of brain glycogen after decapitation (up to 90% in the first minute in adult rats but almost no loss up to 4 min in neonatal animals). Microwave treatment and immersion in liquid nitrogen can substantially reduce glycogen breakdown during the postmortem period. Glycogen is a reserve fuel to be used in an emergency when the blood glucose falls below tolerable levels. This does not mean brain glycogen is metabolically inert; up to 17 μmole/kg/min is metabolized in the rat (Watanabe and Passonneau, 1973). While liver glycogen is designed to provide fuel for the entire body, brain glycogen is used only for local needs. A glycogen-storage disease known as Pompe's disease has been identified. Increased concentration of glycogen in muscle, liver, heart, glial cells, and nuclei of brain stem have been found. Hers (1963) was the first to characterize the enzyme defect (genetic in origin) as the lack of α-1,4-glucosidase in Pompe's disease. The disease manifests nonneural involvement such as a striking enlargement of the heart. Usually the patient lives for only 5–6 months and the cause of death is mostly due to respiratory failure.

In summary, the effects of dietary carbohydrate on the brain are not direct but mediated by hormonal and enzymatic actions, except during acute hypoglycemia. The most common disease, diabetes, due to decreased insulin levels, causes hyperglycemia. This disease affects both the central as well as peripheral nervous system. An untreated diabetic can quickly develop metabolic acidosis and coma. The coma may be due to hyperosmolarity. Thus both hypo- and hyperglycemia can lead to a comatose condition. Dehydration brought about by water deprivation and manitol loading produced metabolic signs similar to diabetic conditions thus supporting the theory of hyperosmolarity and coma in hyperglycemia. However, the question of comatose conditions in the presence of adequate sugar and oxygen in diabetic acidosis remains to be solved. Damage to the peripheral nervous system is also common in diabetics. Although blood insulin cannot cross the blood capillaries, endogenous insulin does occur in cerebrospinal fluid (CSF) (Balazs, 1970) and may influence brain metabolism *in situ.*

6.10. EFFECT OF STARVATION

Many researchers equate lack of carbohydrate with lack of food because, at least in humans, generally most meals contain an appreciable amount of carbohydrate. Deletion of carbohydrates, or a drastic reduction in its intake, has become popular with sections of the population who tend to add body weight easily. There is reason to believe that a diet high in proteins, fats, and cholesterol but very low in carbohydrates can lead to fatty livers (Martins and Dhopeshwarkar, 1981). Total starvation leads to higher levels of ketone bodies in adults along with an increase in FFA levels in the blood plasma. Up to a certain extent, the adult brain adapts to utilize this source of energy in the absence of glucose. In the peripheral tissues, glucose utilization is replaced by FFA oxidation to provide energy. Whatever portion of glucose does enter the peripheral tissues is broken down to lactate and pyruvate (and not all the way to CO_2 and H_2O) which are returned to the liver for resynthesis of glucose. Starvation causes an increase in glycine, a progressive decrease in arginine, but no change in lysine. Alanine also decreases but this amino acid is a preferred substrate for gluconeogenesis in the liver (Cahill, 1970). Thus, many metabolic changes occur during fasting; some of these may be common to the nutritional status during consumption of a so-called, low-carbohydrate (special) diet, particularly when total calories are severely restricted. Due to some adaptive changes, the restricted passage of certain compounds into the brain, and the capacity of the brain to utilize glucose and oxygen even when its concentration is decreased in the blood, the adult brain is protected from ill effects.

Starvation during pregnancy, on the other hand, has very serious effects on the fetus. Recently, a great deal of attention has been focused on this subject. This stems from the fact that there seems to be a tendency toward decreased food intake (mild starvation) by pregnant women who are afraid of permanent obesity. Sometimes this is self-monitored rather than under the competent care of a physician. Most of these women are taking unnecessary chances. They may not be aware of the effects of such a restricted diet on the complex growth starting from a fertilized egg and ending in a multicellular complex organism. One needs only to remember that fetal tissue is built up entirely from nutrients provided by the mother. Starvation can cause hypoglycemia under normal conditions, but in pregnancy, since the fetus is constantly removing blood glucose, starvation produces a much more severe decrease in blood sugar. Additionally, a decrease in glucogenic amino acid levels further decreases gluconeogenesis and this aggravates the problem. Decreased amounts of fuel mean stunted growth with all the accompanying problems. A loss of 10–15% in body weight in a nonpregnant women may be harmless, but the loss of 1 to 2 kg of body weight during pregnancy can be serious to fetal growth (Stein et al., 1975). Animal studies in this regard (Frazer and Huggett, 1970) indicate that although a minimum amount of nutrient is extracted from the mother by the fetus, the remaining extra nutrients are distributed so that about 75% goes to the fetus and the rest goes to the mother. If there is no excess available, some needs of the fetus may not be fulfilled. If by dietary manipulation the rat is not allowed to gain weight during pregnancy, the growth of the fetus is significantly reduced.

The amount of circulating nutrients in the mother is partly governed by insulin action. Although intact insulin cannot be transported across the placental barrier, insulin levels and response to feeding or fasting regulate the maternal blood composition and indirectly the nutrition of the fetus. A mild diabetic condition in pregnancy (fasting blood sugar below 105 mg/dl) shows some characteristic changes in blood composition: FFA levels after an overnight fast tend to be higher and do not decrease after ingestion of a meal, triglyceride values are also higher, and the glucogenic amino acid, serine, remains higher during overnight fasting. Thus, the insulin effect is seen on a much wider scale than just blood glucose (Freinkel and Metzger, 1979). The effect of such a mild diabetic condition on the offspring was seen in increased birth weight. Thus, this mild condition, in fact, mimics the well-established hyperglycemia–hyperinsulinemia syndrome. In this condition, diminished insulin in the mother leads to hyperglycemia, glucose freely gets into the fetal circulation, causes an insulin response in the fetus leading to increased deposition of fat and glycogen, resulting in the formation of large babies. Although it is easier to understand the effects of hyperglycemia, it is difficult to understand the effects of hypertriglyceridemia because triglycerides do not cross the placental barrier. It is possible that they are hydrolyzed slowly to release FFA which can cross the placenta

(Moore and Dhopeshwarkar, 1980). Mortality in babies heavier than normal (~4000 g or over) is mostly due to complications arising from difficulties during delivery; but mortality in babies whose growth is reduced is not only higher, but can be due to multiple causes.

Studies by Drillien (1970) and Wiener (1970) indicate that babies born with low birth weight develop a greater deficit in mental capacity. Birth weight under 1500 g results in brain damage and the subsequent deficit in mental performance. The incidence of cerebral palsy is much higher in this group. Fitzhardinge and Stevens (1972) have summarized that a SGA birth generally does not lead to cerebral palsy but other functions such as fine coordination and attention span were decreased and babies were found to be hyperactive.

7

VITAMIN DEFICIENCIES AND EXCESSES

7.1. INTRODUCTION

The outstanding achievement of nutritional science has been the discovery of about 50 nutrients essential for man. In 1903, F. Gowland Hopkins recognized the need of organic growth factors now known as vitamins. Sixteen different vitamins have been discovered followed by isolation, elucidation of the structure, and in most cases synthesis of these substances. Identification of the role of vitamins as coenzymes was a major step in the understanding of their role in metabolism and physiology. This has been termed a golden era of nutritional research.

There are primary deficiencies of vitamins due to a lack of dietary intake of food containing these vitamins. But many times the deficiency state is brought about by factors other than dietary causes. For example, defects in absorption or storage may be the underlying cause of vitamin deficiency, even though the diet contains adequate amounts. There could be a genetic defect that can bring about underutilization of vitamins by way of enzyme inadequacies and abnormal metabolism. These could, in general, be termed secondary defects.

Gastrointestinal problems such as peptic ulcers, food allergy, intestinal obstruction due to tumors are examples in which absorption, even in the presence of adequate dietary intake, could be suboptimal. Alcoholism or ingestion of "empty calories" can lead to a variety of nutritional deficiencies, including those of the B vitamins. Thiamine (vitamin B_1), for example, is poorly available to alcoholics, particularly if they have liver disease. Poor absorption could be due to a decreased level of dephosphorylation of dietary thiamine phosphate (Baker *et al.*, 1975). In the case of thiamine (vitamin B_1), the presence of thiaminase in food (certain fresh water fish) can lead to thiamine deficiency. Thiaminase displaces the thiazole moiety and this structural change causes the loss of biological activity (Tanphaichitr, 1976). A thermostable antithiamine factor was also found in vegetables and plants and is related to caffeine and tannic acid (Hilker *et al.*, 1971).

Certain drugs, when given for therapeutic purposes, can influence the absorption and metabolism of vitamins (drug–nutrient interaction). The absorption of vitamin B_{12} is impaired by *p*-aminosalicylic acid, colchicine, and neomycin (Yosselson, 1976). Aspirin has been implicated in low serum levels of folic acid in patients with rheumatoid arthritis (Alter *et al.*, 1971). L-DOPA used to treat Parkinson's disease interacts with vitamin B_6, nullifying the beneficial effect of the vitamin (Ellenbogen, 1980).

Malignancy sometimes creates an extra demand for vital nutrients such as vitamins and essential amino and fatty acids, leading to a deficiency state. An increased metabolic rate, hyperthyroidism, can also raise the vitamin requirement and unless the condition is recognized and dietary intake increased, vitamin deficiency will be produced. Any abnormality in absorption caused by vomiting or diarrhea will certainly lead to a deficiency state.

7.1.1. Vitamins as Coenzymes

In general, a coenzyme attaches to an apoenzyme which goes to form an intermediate, holoenzyme. The holoenzyme then combines with a substrate to form a tightly bound enzyme–substrate complex. Under appropriate conditions, such as suitable pH, the enzyme converts the substrate to the product. Although usually the association between coenzyme and apoenzyme is in the form of a covalent linkage, in certain cases, like in lipoic acid and biotin, the vitamin is covalently linked as the prosthetic group of the enzyme itself. In many cases the association may just bring about conformational changes. The binding of the coenzyme and holoenzyme may also influence its metabolism (synthesis and degradation) according to Wyngaarden (1970). There could be a genetic defect in one of the above steps that may result in ineffectiveness of the vitamin as a cofactor and ultimately produce severe deficiency.

7.2. THIAMINE DEFICIENCY

Vitamin B_1, thiamine, is one of the earliest vitamins detected; hence, a lot more information on its biochemical, physiological and nutritional effects is available. Eijkman in 1907 first demonstrated that polyneuritis was induced in pigeons fed polished rice, and the symptoms were reversed by feeding a component that was removed during milling of the rice. The crystalline form was available by 1926 and proved to be biologically active. Classical studies by several workers including Banga *et al.* (1939) demonstrated that during vitamin B_1 deficiency there was an accumulation of lactate in the brain stem and oxidation of pyruvate to acetate was decreased. In 1936, vitamin B_1 was synthesized by Williams (1936). It was soon discovered that thiamine pyrophosphate acted as a coenzyme in the oxidative decarboxylation of pyruvate to acetyl-CoA, a reaction catalyzed by the enzyme pyruvic dehydrogenase. The oxidative decarboxylation of α-ketoglutaric acid, α-ketoglutarate + CoA + $NAD^+ \to$ succinyl-CoA + CO_2 + NADH, is also mediated by thiamine pyrophosphate (TPP). TPP is a component of the α-ketoglutarate dehydrogenase complex. Another decarboxylation reaction similar to the above reaction deals with metabolic intermediates of branched-chain amino acid metabolism. In this reaction TPP is again a required coenzyme.

Leucine $\to$ α-ketoisocaproic acid $\xrightarrow{\text{TPP}}$ isovaleryl-CoA

Isoleucine $\to$ α-keto-β-methylvaleric acid $\xrightarrow{\text{TPP}}$ α-keto-β-methylbutyryl-CoA

Valine $\to$ α-ketoisovaleric acid $\xrightarrow{\text{TPP}}$ isobutyryl-CoA

However, no specific accumulation of keto acids or branched-chain amino acids has been shown to occur in thiamine deficiency.

A further characterization of thiamine brought about new information regarding the various active forms. Approximately 80% of thiamine was found to be in the form of TPP and another 10% occurs as the monophosphate ester (Fig. 15). Thiamine is found in all tissues, but heart, kidney, liver, brain, and muscle seem to have a higher amount. The susceptibility of the brain could be due to its higher concentration in this tissue. Another enzyme with which thiamine acts as a coenzyme is transketolase which catalyzes reactions involved in direct oxidation of glucose by the pentose shunt pathway (Horecker and Smyrniotis, 1953) (Fig. 16). The pentose shunt pathway is more predominant in the developing brain and thus the requirement of thiamine may also be increased during this period. The levels of the enzymes that require

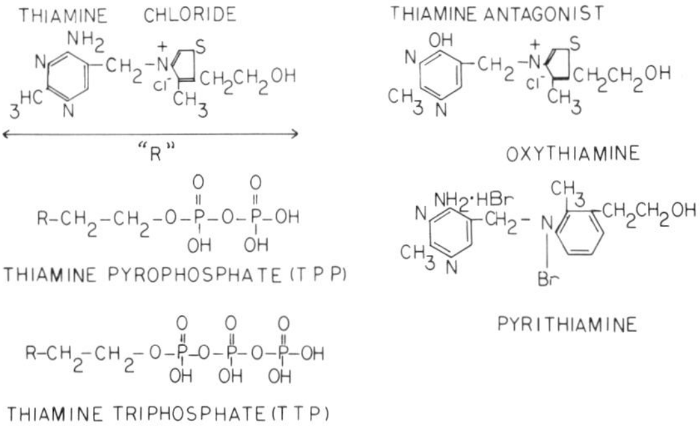

Figure 15. Thiamine, phosphate esters, and thiamine antagonists.

thiamine, pyruvate dehydrogenase, and transketolase have a different pattern in the various areas of the brain (McCandles and Schenker, 1968).

When rats were put on a thiamine-deficient diet, the level of vitamin B_1 was reduced by about 50%; but, at this stage, no obvious growth retardation was observed. A further depletion to about 70% of the control value resulted in unsteady gait and slow movement (Dreyfus and Victor, 1961). Therefore, the rat is not a preferred test animal for studying thiamine deficiency because levels have to fall below 20% of normal before any classic signs such as disturbance in posture and equilibrium are observed. To overcome this difficulty, a thiamine analog, pyrithiamine, is included in the diet. This compound inhibits the utilization of thiamine and, of particular interest, exerts its effect on the CNS (Wooley and White, 1943).

Rats raised on a vitamin B_1-deficient diet show signs such as ataxia and loss of the righting reflex. It is interesting to note that TPP decreases much more than other derivatives of thiamine. Since pyruvate to acetate is a major metabolic reaction, mediated by TPP, which provides acetyl-

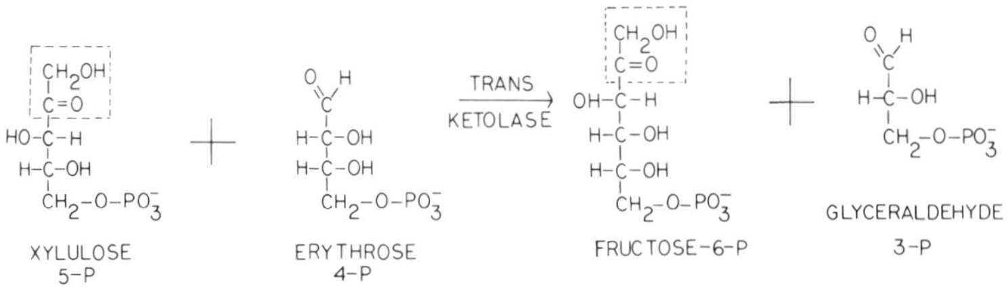

Figure 16. Transketolase reaction (dotted lines show group transfer).

CoA for fatty acid synthesis, one would have expected some kind of disturbance in lipid metabolism or myelin deposition but no such specific effect has been observed in the immediate postnatal period. However, if the deficiency is instituted during pregnancy, reduction in fetal weights has been observed. Gangliosides were increased but other sphingolipids were decreased (Geel and Dreyfus, 1975). Similarly, the pentose phosphate pathway depends on transketolase activity which, in turn, is reduced in thiamine deficiency. The reduced activity of the pentose shunt pathway would also decrease production of NADPH; and, it is well known that the availability of NADPH regulates lipogenesis. However, no alterations in the fatty acid profiles have been observed in brain fatty acids in thiamine-deficient rats (McCandles and Schenker, 1968).

The role of thiamine in nerve stimulation has recently been studied by Itokawa *et al.* (1972). These authors indicate that thiamine is released when spinal or peripheral nerves are electrically stimulated. Pyrithiamine seems to displace thiamine from the active site, and no enzymes requiring TPP were inhibited; thus, the action was directly on thiamine rather than via thiamine-dependent enzymes. This new role assigned to thiamine in nerve stimulation also indicates that there are thiamine receptive sites in the neuronal membrane and this may explain certain neurological symptoms in thiamine deficiency. Tanka and Cooper (1968) showed that thiamine is present only in nerve membranes and not in axoplasm. Recently, it was found that brains of patients who died of subacute necrotizing encephalomyelopathy contained either a very small amount, or no detectable amount of TPP. Further, urine, blood, and CSF of these patients contained a protein factor that inhibited *in vitro* synthesis of thiamine triphosphate from thiamine diphosphate (Pincus, 1972).

Thiamine deficiency in chronic alcoholics (Wernicke–Korsakoff syndrome) was noticed early as an entity related to poor nutritional status of all alcoholics. The symptoms of the Wernicke–Korsakoff syndrome, however, were noted in both chronic alcoholics and nonalcoholics (Victor *et al.*, 1971). Mental symptoms are present in over 90% of the cases, starting with delirium. Past memories are lost and, in acute cases, eye movement is affected. The level of transketolase in the RBC is decreased in patients suffering from this syndrome and thiamine treatment promptly restores the enzyme activity. Pyruvate levels in the blood are elevated in acute stages of the disease due to decreased oxidative decarboxylation to acetate (TPP is essential for this enzyme activity) (Victor *et al.*, 1957). It has been shown that nystagmus and opthalmoplegia and, to a lesser extent, ataxia of the Wernicke–Korsakoff syndrome could be reversed within hours or days by oral intake of thiamine even in patients who continue to drink alcohol. But mental confusion, learning disability, and loss of recent memory cannot be restored completely; these conditions may have been the result of irreversible pathological damage.

Infantile beriberi, characterized by anorexia, vomiting, and edema, has been recognized for many years (Hirota, 1898) and is reversed by thiamine administration. Due to early recognition of this disease, cases of infantile beriberi are not commonly seen even in poorer countries. In fact, the polishing of rice, a method that was commonly followed, has long been recognized to remove the B-vitamins and is no longer followed. If for aesthetic reasons polished rice is desirable, other dietary items can be easily fortified (e.g., bread) to overcome the nutritional loss.

Leigh's disease, a disease of genetic origin in infants, is correlated to a lack of triphosphothiamine in the brain (Pincus, 1972). An inhibitor of the enzyme that synthesizes triphosphothiamine is thought to be circulating in the blood but has not yet been characterized. In another variation of this disease, pyruvic decarboxylase was found to be effective (Blass *et al.*, 1970). Oral administration of thiamine and certain dietary restrictions may prove to be beneficial. A patient was identified with a variant of Maple Syrup Urine disease, where the levels of blood amino acids were increased, and started on a daily dose of 10 mg of thiamine. This produced a complete reversal of the aminoaciduria (Scriver *et al.*, 1971).

Warnock and Burkhalter (1968) have ascribed to thiamine a new and different role in the physiology of the brain. They argue that up until now the biochemical lesion of vitamin B_1 deficiency is defined as a decreased ability to decarboxylate pyruvate, which, in the brain, would lead to polyneuritis. But other investigators have shown that this defect occurs only in pigeons (Koeppe *et al.*, 1964); thus, Warnock and Burkhalter think that some other mechanism must be involved in thiamine deficiency in rats. The authors suggest that the blood–brain barrier may be malfunctioning in thiamine deficiency.

Pyruvate can be converted to glutamate by two pathways. One is a direct uptake leading to formation of [5-^{14}C]glutamic acid and the other pathway in which the labeled pyruvate is first converted to glucose which then enters the brain, is degraded again to pyruvate, and forms glutamic acid. However, the labeling pattern of glutamic acid by the two different pathways is different as shown in Fig. 17. Using this rationale, [2-^{14}C]pyruvate was injected *iv*, and 10 min later the animals were sacrificed. Examination of brain glutamate (particularly the radioactivity in carbon 4) indicated that in adult rats [2-^{14}C]pyruvate entered the brain directly. But Koeppe and Hahn (1962) had already shown that pyruvate

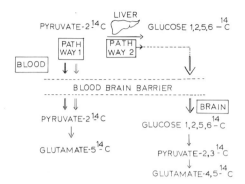

Figure 17. Schematic representation of pyruvic acid transport into the brain. Adapted from Warnock and Burkhalter (1968).

does not enter the normal brain of adult animals. Thus, it seems that vitamin B_1 deficiency caused some alteration in the blood–brain barrier system that led to the entry of the pyruvate. The alteration of the barrier system was very subtle and gentle enough that trypan blue was still prevented from diffusing out of brain capillaries. Recent studies by Oldendorf have indicated that pyruvate is not excluded from entering the brain following intracarotid injection. In fact, the index of permeability for pyruvate was 24. When compared with the reference standard ($3H_2O$), arbitrarily chosen to have an index of 100. Compared to glutamic acid that had an index of 3 and known to be restricted by the barrier, pyruvate seems to penetrate normal brain much more readily. In light of this observation, it is difficult to interpret Warnock and Burkhalter's results.

7.3. VITAMIN B_2 (RIBOFLAVIN) DEFICIENCY

The role of riboflavin in cellular metabolism is dependent on its conversion to the coenzyme form (Fig. 18). Riboflavin 5-phosphate, commonly referred to as flavine mononucleotide (FMN) and riboflavin 5-pyrophosphate adenosine, referred to as flavine adenine dinucleotide (FAD), are the two coenzyme forms of riboflavin (Abraham, 1939). The coenzymes serve as carriers in the electron transport systems leading to the formation of the high-energy phosphate bonds of ATP. FMN is a part of the L-amino oxidase system that is involved in oxidation of L-α-amino acids and L-α-hydroxy acids to α-keto acids. FAD is a part of glycine oxidase, glucose oxidase, diamine oxidase, and xanthine oxidase.

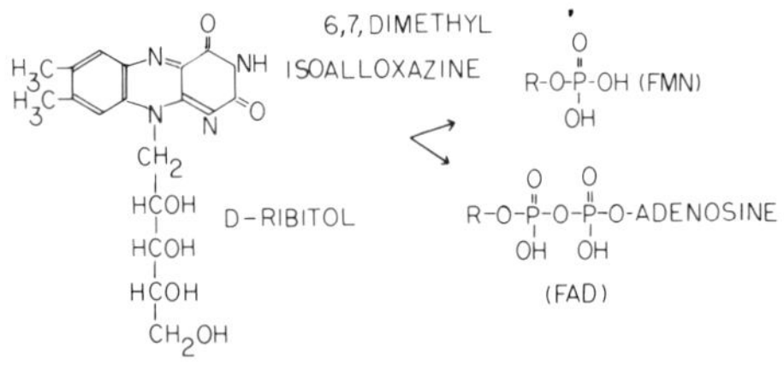

Figure 18. Riboflavin and its coenzyme forms.

The acyl-CoA dehydrogenases are flavoproteins containing FAD and are responsible for introducing double bonds in the process of β-oxidation of fatty acids.

It is easy to produce dietary deficiency of riboflavin resulting in cessation of growth and ultimately death. Even in moderate deficiency, although the liver loses 60% of its riboflavin, the loss in brain tissue is only marginal (Burch *et al.*, 1956). Riboflavin deficiency produces neurological symptoms, such as partial paralysis of the hind legs in monkeys (Mann *et al.*, 1952) possibly due to degeneration of myelin in the peripheral nerves. Since riboflavin plays a part in monomine oxidase, it is possible that riboflavin deficiency can cause a disturbance in the metabolism of neurotransmitters, but no such data are yet available. Cow's milk contains five times as much riboflavin as human milk, but no deficiency has been noticed in infants fed solely on breast milk. In fact, frank riboflavin deficiency in humans is rarely seen. So far no specific ill effects of riboflavin deficiency can be related to brain biochemistry or function.

7.4. NICOTINIC ACID (NIACIN) DEFICIENCY

For many years pellagra was characterized as a disease of dietary deficiency. It was believed that this disease was common to populations that had corn as a major staple food. The disease was endemic in the southern part of the United States and Dr. Goldberger, from 1914, undertook an extensive study to determine the cause of this disease. He was first to point out that the disease could not have been due to an infectious agent. He found that although inmates of an institution and the staff lived in the same environmental surroundings, only the inmates showed signs of the disease. The reason the staff did not have this disease was further traced to their diet, particularly to the inclusion of animal protein in their meals. Finally, in 1928, by using diets that by that time were believed to produce pellagra in humans, Goldberger and Wheeler produced a condition known as "black tongue" in dogs, very similar to human pellagra. Elvehjem *et al.* (1938) isolated nicotinamide as the pellagra-preventive factor.

Nicotinamide is the physiologically active form of niacin (Fig. 19). Nicotinamide functions as a component of a coenzyme, just as do the other B vitamins. The coenzyme nicotinamide adenine dinucleotide (NAD) and its phosphate ester, NADP, were previously known as coenzyme I and coenzyme II or DPN and TPN. The coenzymes are involved in electron transport in cellular respiration. Since this is a basic cellular process, lack of this vitamin (and in turn the coenzymes) affects cellular survival. Reduced coenzyme NADH donates hydrogen to FAD but NADPH gives up its hydrogen in synthetic reactions of fatty acid synthesis and steroid hydroxylations.

It was known that pellagra was not observed in persons who included animal proteins in their diet. The explanation comes from the discovery that dietary tryptophan was a precursor of nicotinic acid. Corn was found to be deficient in this amino acid and thus led to a deficiency of nicotinic acid.

Tryptophan $\xrightarrow[\text{pyrrolase}]{\text{Try}}$ formylkynurenine $\xrightarrow[\text{formylase}]{\text{kynurenine}}$ kynurenine $\xrightarrow[\text{hydroxylase}]{\text{kynurenine}}$ 3(OH)-kynurenine

$\xrightarrow{\text{kynurenase}}$ 3(OH)-anthranilic acid $\longrightarrow$ quinolinic acid $\longrightarrow$ nicotinic acid

It is now believed that approximately 60 mg of tryptophan is equivalent to 1 mg of nicotinic acid. Although this conversion takes place readily in the liver, the brain lacks the necessary enzymes (Ikeda *et al.*, 1965). Although nicotinic acid is not readily taken up by the brain following intraperitoneal injection, intracisternally-injected nicotinic acid is incorporated into brain NAD. In further experiments, Deguchi *et al.* (1968) found that niacin ribonucleoside was a better precursor, thus suggesting the riboside as the transport form of niacin in the brain.

Frank niacin deficiency in humans may arise not only from a lack of tryptophan and niacin in the diet (such as in a high-corn diet) but because of an imbalance in amino acid content of the diet (Gopalan and Shrikantia, 1960). At the same time, one can obtain nicotinic acid from

NICOTINIC ACID   NICOTINAMIDE

Figure 19

trigonelline found in coffee beans. Nicotinic acid, being stable to heat and light, is not lost during food processing, including the roasting of coffee beans.

Dermatitis, diarrhea, and dementia (the well-known "3 D's") are seen in pellagra; but in a milder form of the disease, depression, emotional instability, and impairment of memory are also common (Jolliffe *et al.*, 1940). Blood vessels in the CNS show fatty degeneration, and areas of myelin seem to be degenerated with free fat in the nerve trunk (Langworthy, 1931). A deficiency of niacin also causes the disappearance of Nissl bodies.

A genetic defect, due to an inherited autosomal trait, causes a disease known as Hartnup's disease. The clinical symptoms of this disease resemble those of pellagra, including neurological involvement. Some patients are mentally retarded (Jepson, 1972). In this disease, there is a defect in the intestinal and renal transport of neutral amino acid. Although this includes several amino acids, such as leucine, isoleucine, tyrosine, and tryptophan, the symptoms seem to be related to poor tryptophan absorption. The pellagra-like symptoms of Hartnup's disease (neurological and dermal) respond to treatment with nicotinamide.

The biochemical lesion of pellagra can be studied in animal models. In rats raised on a tryptophan-containing (niacin-free) diet there was significant fall in brain DNA and NADH levels and a lesser decrease in NADP and NADPH (Garcia-Bunuel *et al.*, 1962). Since 6-aminonicotin-amide is an antimetabolite of niacin, this compound has been used to produce niacin deficiency. A single dose resulted in paralysis. Chronic treatment with small doses causes neurological degeneration even in adult animals (Sternberg and Phillips, 1958). There is a competitive inhibition of several NADP-dependent enzymes, causing a decrease in the HMP shunt pathway. Since the developing brain uses this pathway (in order to make available NADPH needed for lipid synthesis and myelination), niacin deficiency during the early growth period may lead to myelination defects.

A new physiological role for niacin has been recently suggested in counteracting the effects of Paraquat [1,1'-dimethyl-4,4'-bipyridinium (cation) dichloride], a nonspecific herbicide used commercially in agriculture that is toxic to rodents as well as to man under certain conditions.

7.5. VITAMIN B₆ (PYRIDOXINE) DEFICIENCY

Although pyridoxine was isolated in 1938 independently by Kerestezy, Gyorgy, and Lepkovsky, it was not until 1945 that Snell identified the three different forms of the vitamin (Fig. 20). Lack of this vitamin in the rat brought about loss of hair, alopecia, and a dermatitis characterized by scaliness about the paws and mouth. Frequent seizures are observed in pyridoxine-deficient newborn rats.

Like other members of the B vitamins discussed above, this vitamin also forms a coenzyme, pyridoxal phosphate. The coenzyme is involved in reactions such as transamination, deamination, and desulfhydration of amino acids and various other decarboxylation reactions. There are more than 50 such enzymatic reactions in which vitamin B_6 is involved. The reactions, in addition to the requirement for the coenzyme, also need metal ions such as Fe^{3+}, Cu^{2+}, and Al^{3+}. Pyridoxine has been implicated in the conversion of linoleate to arachidonate (Witten and Holman, 1952). However, the desaturation and/or elongation steps involved in this type of transformation seem to be independent of this cofactor requirement and thus this nutritional finding and its implication is not clear. A combination of phosphokinase, transaminase, and an oxidase catalyze the transformation of pyridoxine, pyridoxal, and pyridoxamine into their phosphate esters and their interconversions. Pyridoxal phosphate and pyridoxamine phosphate account for most of the naturally occurring forms in the tissues. In the rat brain the level of pyridoxal phosphate rises with age starting from birth. By 30 days, 80% of the adult level is already reached (Loo, 1972). Ebady et al. (1968) found that the higher the concentration of biogenic amines, the lower the activity of pyridoxal phosphokinase. 5HT seems to be more effective in this respect. Although the liver and muscle readily incorporated injected radioactive pyridoxine, the uptake in the brain was delayed. The incorporated pyridoxine was ultimately converted to pyridoxamine 5-phosphate in the brain in situ (Tiselius, 1973).

Brain tissue is rich in glutamic acid; in fact, the concentration of free glutamic acid is 5 times higher than that of aspartic acid, 10 times higher than that of serine, and more than 150 times greater than that of tyrosine. One of the major metabolic alterations of glutamic acid, apart from the formation of glutamine, is decarboxylation to form GABA and this reaction is uniquely confined to the nervous system. GABA is further metabolized to succinic acid.

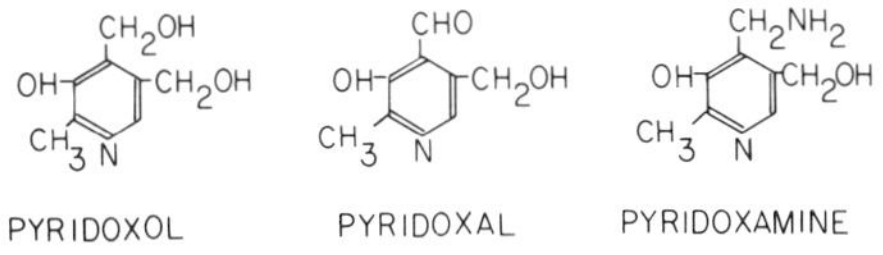

Figure 20.   The three forms of vitamin B_6

From the scheme outlined in Fig. 21 it is quite apparent that vitamin B_6 plays a key role in brain metabolism. GAD has been shown to be more sensitive to pyridoxine deficiency than is the transaminase (Massieu *et al.*, 1962). Higgins (1962) found that rats fed a pyridoxine-deficient diet for 4 weeks showed a decreased amount of GABA in the brain. Tews (1969) found that after 4 weeks on a pyridoxine-deficient diet, levels of GABA, serine, and alanine were significantly decreased in mouse brain. He argued that glucose is converted to serine via the phosphohydroxypyruvate-to-phosphoserine pathway. This reaction is catalyzed by a transaminase that requires pyridoxal phosphate. Thus, in vitamin B_6 deficiency, the levels of serine would also be affected. Experimental rats raised on a pyridoxine-deficient diet seemed to show decreased immunity; and, if the stress was instituted early enough, it may cause permanent immunological defects in the fetus that could become manifest in later life (Davis *et al.*, 1970). Bayoumi and Smith (1972) have reported that vitamin B_6 deficiency during the terminal period of pregnancy caused a decrease in brain weight of newborn animals and the growth retardation continued if the deficiency was not corrected during the next 3 weeks.

Sphingosine is a component of sphingoglycolipids. The biochemical pathway for the synthesis of sphingosine starts with palmitic acid and serine. The decarboxylation of serine which is one of the reactions in this pathway depends on pyridoxine (Fig. 22).

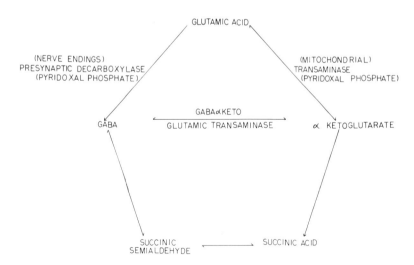

Figure 21.   Metabolic pathways of GABA.

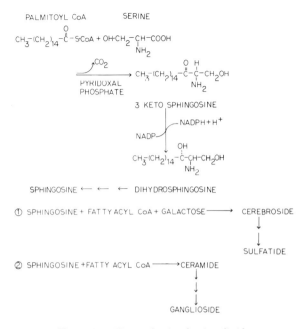

Figure 22. Biosynthesis of spingolipids.

Thus, lack of vitamin B_6 can be expected to cause a decreased synthesis of cerebroside and sulfatides, both of which are important myelin lipids. The gangliosides are found in higher amounts in the neuron-rich gray matter rather than in the myelin-rich white matter of the brain. The cerebral sphingolipids are significantly decreased in vitamin B_6 deficiency (Kurtz *et al.*, 1972). Other decarboxylation reactions that require vitamin B_6 as a cofactor deal with (1) decarboxylation of 5(OH)-tryptophan to 5HT and (2) decarboxylation of DOPA to dopamine. It is conceivable that in pyridoxine deficiency the levels of 5HT and dopamine could be reduced. However, although the level of the L-aromatic amino acid decarboxylase enzyme is reduced in vitamin B_6 deficiency, the levels of the NA and dopamine neurotransmitters are apparently not affected significantly (Le Blancq and Dakshinamurty, 1975). Other enzymes that are dependent on vitamin B_6 are (1) cystathionase (cystathionine → serine + homocysteine), (2) cysteic acid decarboxylase (cysteic acid → taurine + CO_2), and (3) kynureninase (kynurenine → anthranilate + alanine). The enzymes cystathionase and cysteic acid decarboxylase may be important in the brain considering that cystathionine and taurine are present in high concentration, at least in rat brain. The third enzyme may be important for the conversion of tryptophan to niacin (but this conversion is, at best, a weak reaction in the brain).

Some symptoms arising out of drug use are believed to be responsive to pyridoxine but the mechanism remains unknown. In uremic patients low levels of plasma pyridoxine and glutamic–oxaloacetic transaminase activity in RBC suggests pyridoxine deficiency (Stone et al., 1975). The exact cause of this nondietary deficiency in uremics is not known. Pyridoxine deficiency has been suspected in tuberculosis patients undergoing isoniazid treatment (Standal et al., 1972).

Pyridoxine deficiency in infants warrants special mention. Some infants show pyridoxine dependency, i.e., doses in the range of 5–25 mg are needed to control seizures. If the condition is not diagnosed early, severe mental retardation may set in. Infant formulas are now required to be fortified with vitamin B_6 after a disastrous earlier incident, when a formula produced convulsions. That preparation had been subjected to prolonged heat treatment as a means of reducing the allergic properties of cow's milk (Tomarelli et al., 1952).

7.6. FOLIC ACID DEFICIENCY

Folic acid, folacin, and pteroylglutamic acid are synonyms for this vitamin. Its discovery originates in symptoms observed in people suffering from anemia whose diet consisted primarily of white rice and bread. The anemia responded to yeast. The unidentified factor was termed Wills factor, citrovorum factor, etc. Mitchell and co-workers (1941) isolated a compound from spinach which was found to be necessary for the growth of *Streptococcus faecalis* R. and named it folic acid (Fig. 23).

Pteroylglutamic acid is a vitamin for most mammals whereas p-aminobenzoic acid was found to be essential for some bacteria that can

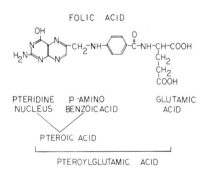

Figure 23. Components of folic acid.

synthesize the larger molecule. The growth factor essential for certain bacteria and a factor isolated from yeast and liver extracts were all found to be able to cure anemia produced in monkeys. The active principle was finally isolated and the structure characterized by chemical methods in 1948 (Mowat *et al.*, 1948).

The richest source of folic acid in a vegetarian diet is leafy vegetables, and for nonvegetarians, liver is a very good source. Yeast also contains an appreciable amount of this vitamin. But heating (normal cooking) at 100–120°C for about 10 min brings about a ~50–60% loss. Most of the foods contain folic acid not as a monoglutamate but as a polyglutamate. This means the glutamic acid residues need to be split to form monoglutamate before absorption in the gastrointestinal (G.I.) tract. Although the gut bacteria can synthesize the vitamin, this may not be enough to meet daily requirements.

The active form of this vitamin is tetrahydrofolic acid (FH_4) and is formed from the dihydro form, FH_2, by a reductase with NADH or NADPH serving as carriers. Ascorbic acid may be helpful in this reduction. FH_4 is a carrier of a "one-carbon" moiety, e.g., –CHO (formyl) or –CH_2OH (hydroxymethyl), and is comparable to CoA for the two-carbon moiety, acetate. The single-carbon units are important in the synthesis of purines and pyrimidines. Some examples are the conversion of glycine to serine and homocysteine to methionine, methylation of a pyrimidine intermediate to thymine, and introduction of carbon 2 and 8 into the purine ring structure (Phear and Greenberg, 1957). Its involvement in the formation of nucleic acid gives this vitamin a very basic role in the physiology of the cell. Antagonists such as aminopterine, methotrexate, and 5-fluorouracil are all inhibitors of dihydrofolate reductase and can stop cell division, and, as such, these compounds have been used in cancer therapy.

There is evidence to support the idea that the serum concentration of folate can influence its concentration in the brain or CSF (Reynolds *et al.*, 1972). Injected radioactive folate was taken up by the brain (Levitt *et al.*, 1971). Whole brain folate decreases in mice during maturation. Folic acid deficiency during pregnancy would have far-reaching effects due to its central role in nucleic acid metabolism and thus would affect DNA replication and mitotic activity. In fact, in rats a deficiency of folic acid midway during gestation will result in failure of fetal growth, and fetal resorption occurs. If remedial folic acid is given at this point, one can prevent fetal resorption but the term fetus may have hydrocephalus (Stempak, 1965). Arakawa and co-workers (1969) noted that folic acid deficiency, brought about by dietary inhibition of folic acid, produced defects in basic electroencephalograph (EEG) patterns in young rats. If a similar inhibitory procedure is started during gestation, microcephaly is observed in newborn rats (Arakawa *et al.*, 1967). Decreased levels of hydroxy fatty acid (myelin lipids) have been noted in folic acid deficiency (Chida *et al.*, 1972), which also has been shown to produce learning deficiencies in animals (Whitley *et al.*, 1951). Not all ill effects on the CNS are due to a deficiency of folate; even an excess of folate can be damaging as proved by intraventricular injection or peripheral injection of sodium folate (Hommes *et al.*, 1973). Folic acid deficiency has also been noted in clinical studies. The clinical features include mental retardation and abnormalities in EEGs. Arakawa (1970) in a study reported four patients with inborn errors of metabolism resulting in deficiencies of two enzymes, formiminotransferase and N-methyltetrahydrofolate transferase. Both of these enzymes can impair nucleic acid biosynthesis and bring about neurological damage. Folic acid sometimes used therapeutically in pernicious anemia may bring about a reversal of hematologic abnormalities but it does not correct symptoms associated with the nervous system. In fact, it may be risky to use this vitamin in pediatric patients (Holt, 1972). Even in adults, excess folic acid proved to be harmful. Hunter *et al.* (1970) found that 15 mg daily for about a month caused mental disturbance, changes in sleep patterns, and G.I. effects in normal adult volunteers.

Biochemical reactions leading to biosynthesis of methyl groups of methionine are dependent on folic acid. But when a drug such as L-DOPA is administered it was found that folic acid-deficient rats could not maintain normal levels of s-adenosylmethionine (Ordonez and Wurtman, 1974). When L-DOPA, a precursor of dopamine, is given (Parkinson's disease) the O-methylation reaction produces O-methyl-DOPA as a major product. Thus, the level of the methyl donor, SAM, decreases (Wurtman *et al.*, 1970). The maintenance of adequate levels of SAM is important for normal functions of the brain. Normal catabolic reactions of catecholamine neurotransmitters involve O-methylation by catechol-O-methyltransferase (COMT). Thus, folic acid deficiency can lead to disturbances in neural transmission.

7.6.1. Drug Treatment and Folic Acid Requirement

Oral contraceptives interfere in the utilization of certain vitamins like pyridoxine and folic acid (Theuer, 1972). Thus, when long-term use is involved, intake of vitamins such as folic acid should be increased to avoid conditions such as folate-responsive megaloblastic anemia (Beutler, 1972). More importantly, the folic acid levels will further decrease during pregnancy (increased need), particularly if before pregnancy oral contraceptives were used for a long time without any additional folic acid supplementation. Another frequently noted observation deals with the use of anticonvulsant drugs; these drugs lead to folic acid deficiency (Reynolds, 1968). Almost 50% of the epileptics receiving anticonvulsant drugs have megaloblastic anemia and over 70% have low folic acid levels in the blood (Waxman et al., 1970). These drugs include diphenylhydantoin, phenobarbital, and pyrimidone, alone, or in combination with other drugs. Anticonvulsant drugs are teratogenic and, therefore, dangerous during pregnancy and the underlying cause may well be folic acid deficiency (Roe, 1974).

Sometimes antimalarials, used frequently in many tropical countries, cause a deficiency of both vitamin B_{12} and folic acid; and if folic acid is given alone, the deficiency of B_{12} may be actually increased. This may be true with the use of anticonvulsants. In these patients many neurological symptoms are observed, e.g., slowing of mental functions, apathy, and dementia.

The correlation between folic acid status and neurological symptoms attracted attention of researchers and physicians toward the biology of schizophrenia in relation to folic acid and vitamin B_{12} (McDonald and Bolman, 1975; Wyatt et al., 1971). The proposed hypothesis considers transmethylation as a pivotal reaction leading to derivatives of biogenic amines, a process in which methyltetrahydrofolic acid plays a central role.

7.7.   VITAMIN B_{12} DEFICIENCY

Almost 60 years ago, Minot and Murphy (1926) reported that inclusion of liver in the diet reversed the pathology of pernicious anemia. The active principle was later identified as vitamin B_{12} (Hodgkin et al., 1955). It was found that this vitamin occurs in several biological active forms. The cyanocobalamine contains a cyanide group attached to the central cobalt atom and it is the major form of vitamin B_{12} in nature. Castle (1929) had observed, long before crystalline vitamin B_{12} was available, that in order to absorb the antipernicious anemia factor (vitamin B_{12}) it was necessary to have an intrinsic factor in the gut; in the absence of such a factor vitamin B_{12} deficiency could be brought about by either low dietary intake (pure vegetarian diet) or by lack of intrinsic factor (genetic), a glycoprotein secreted by the parietal cells of the stomach. Pure vegetarians had a very low blood vitamin B_{12} level (9–17 ng/100 ml) as compared to nonvegetarians living in Bombay (Dhopeshwarkar et al., 1956). The low value was purely due to diet and not due to a lack of intrinsic factor. The binding of this factor and the vitamins from the diet are an essential step in absorption. It has been suggested that the intrinsic factor stimulates pinocytosis, explaining the absorption of a large complex such as vitamin B_{12}–intrinsic factor.

Vitamin B_{12} is a coenzyme required for some very essential biological reactions (see Fig. 24). The two forms of the coenzyme are methylated cobalamine and 5'-deoxyadenosyl derivative. The methylated derivative participates in the transfer of the methyl group from 5-methyltetrahydrofolate to homocysteine to form methionine and tetrahydrofolate (Taylor and Weissbach, 1969).

In the absence of vitamin B_{12}, methylmalonyl-CoA accumulates and is excreted in the urine. The deoxyl derivative of vitamin B_{12} is found in liver, kidney, brain, and CSF. Vitamin B_{12} is found in a relatively higher concentration in human brain choroid plexus compared to white matter. A higher concentration of methylmalonyl-CoA can substitute for malonyl-CoA in the elongation of fatty acids. This will result in the formation of branched-chain fatty acids that may become incorporated into cellular membranes of the brain. Proof of this hypothesis came from the work of Kishimoto *et al.* (1973), who found a 6 to 13-fold increase in both odd-numbered and branched-chain fatty acids in the brain of a patient who died from methylmalonic aciduria. Levy and associates (1970) studied an infant who died at $7\frac{1}{2}$ weeks of age and found an abnormally large amount of cystathionine and homocysteine in blood and urine. Enzymatic analysis of liver, kidney, and brain obtained at autopsy revealed specifically deficient activity in the B_{12}-dependent enzyme N^5-methyltetrahydrofolate homocysteine methyltransferase. Also, the deoxy-

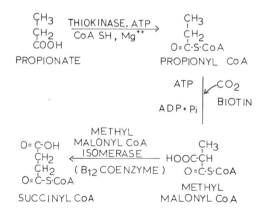

Figure 24. Biochemical action of vitamin B_{12}.

B_{12} levels were very low in the brain but not as low in the plasma or the liver. Thus, it appears to be a defect in the formation of the deoxy coenzyme form of vitamin B_{12} in the CNS.

Methylmalonyl-CoA can also markedly inhibit fatty acid synthetase activity (Frankel *et al.*, 1973). Pfeifer and Lewis (1979) studied the effects on rats of a diet low in vitamin B_{12}, fed for 20 weeks. They found smaller amounts of 20:4ω6 and 22:5ω6 in cerebral EPG, suggesting that vitamin B_{12} deprivation may interfere in the conversion of linoleate to longer-chain PUFA.

Genetic error, resulting in a defect in transformation of vitamin B_{12} to the coenzyme forms, has reportedly resulted in the death of an infant at age 2. In another case, severe mental retardation was noted in a girl who died at age 7. Postmortem examination revealed cerebral atrophy and histological changes similar to those seen in pernicious anemia (Dillon, 1974).

It is conceivable that methionine levels in the brain are lowered in the absence of vitamin B_{12}. The change in concentration would lead to a decreased flux of amino acid incorporation into brain proteins. This hypothesis was supported by the work of Gandy *et al.* (1973), who administered 1-aminocyclopentane carboxylic acid (which is a powerful inhibitor of the homocystein → methionine reaction) and found severe functional abnormalities. The mice showed ataxia, paralysis, and demyelination of the spinal cord.

In summary, vitamin B_{12} deficiency produces severe progressive deterioration in the CNS. These include patchy, diffuse myelination; this may begin in the peripheral nervous system but progresses ultimately to involve the CNS.

7.8. BIOTIN

Rats fed large amounts of raw egg white developed eczema-like dermatitis, spectacle eye (a characteristic loss of hair around the eyes), and sometimes paralysis of the hindlegs. However, if the eggs were cooked this toxicity disappeared. A protective factor present in the liver and yeast was soon isolated that could cure the above condition in the rat. This factor known as *anti-egg white injury* factor was soon isolated and characterized to be chemically identical to biotin (Fig. 25).

Biotin is needed in the "fixation" of CO_2. This role was seen in the conversion of acetyl-CoA to malonyl-CoA and pyruvate to oxaloacetate. The vitamin acts as a coenzyme for the carboxylases that are vital to reactions of intermediary metabolism. For example, acetyl-CoA carbox-

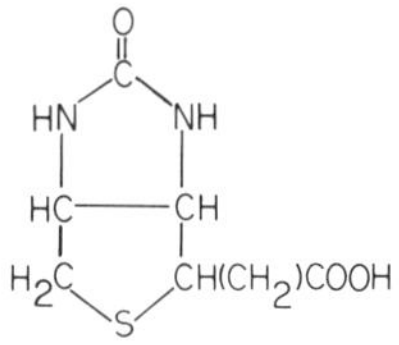

Figure 25. Biotin.

ylase was the first enzyme recognized as a biotin enzyme. Biotin was found to be convalently bound to the enzyme. The overall reaction involves two steps:

biotin–protein + ATP + HCO_3^- → CO_2 ~ biotin–protein + ADP + P_i

CO_2 ~ biotin–protein + acetyl-CoA → biotin–protein + malonyl-CoA

Sum:	ATP + HCO_3^- + acetyl-CoA → ADP + P_i + malonyl-CoA

The importance of this reaction was realized in the overall fatty acid synthesis. It also helped to explain the requirement for bicarbonate in fatty acid-synthesizing systems. It is, in general, considered a rate-limiting step in the synthesis of fatty acids. Similarly, pyruvate carboxylase is vital in carbohydrate metabolism, and propionyl carboxylase is vital to animals that derive their energy from propionic acid. Avidin, a protein found in raw egg white, binds strongly with biotin, thus making it unavailable. Cooking of eggs denatures avidin and thus it cannot interfere in the absorption of this vitamin. Recently, it was found that the biotin-avidin complex when injected intraperitoneally is slowly degraded, making the biotin available again (Lee *et al.*, 1973).

It is hard to produce biotin deficiency because the small amount needed by any species (including humans) can be made in the gut by the microbial flora. Feeding raw egg white may be one of the ways to produce biotin deficiency in experimental animals. Such a deficiency has been studied in puppies (Smith and Lasater, 1945) and pigs (Lehrer *et al.*, 1952). The deficiency caused dermatitis and paralysis of the hindlegs. In the rat, actual lesions of the peripheral nervous system were recorded. Further studies showed that although escape learning was not affected in rats, avoidance learning was impaired in biotin deficiency (Stewart *et al.*, 1966). Sydenstricker (1942) fed human volunteers egg white for over 9 weeks; all subjects showed mild depression, extreme lassitude, and muscle pain. Supplements of biotin cured the symptoms.

Inborn errors of metabolism may affect the various carboxylase enzymes singly or in multiple form. For example, a 2-year-old child with a defect in propionyl-CoA carboxylase was successfully treated with biotin to reduce blood levels of propionic acid (Barnes *et al.*, 1970). β-Methylcrotonylglycinuria and β-hydroxyisovaleric aciduria in a 5-month-old infant were treated with biotin. This brought about relief in abnormal behavior, and the urinary excretion was restored to normal metabolites.

Since biotin is an essential coenzyme in the formation of malonyl-CoA and eventually fatty acids, biotin deficiency can conceivably affect lipogenesis in the brain. If the timing of the deficiency concurs with the growth spurt, myelination could be affected. However, so far this has not been clearly shown with animal models. The denaturation of avidin during cooking may not be complete (e.g., in soft-boiled eggs) and so it is possible that biotin becomes unavailable for intestinal absorption under these conditions. However, humans are also able to obtain adequate biotin supplies via synthesis by the gut flora.

7.9. PANTOTHENIC ACID DEFICIENCY

Pantothenic acid is a constituent of CoA, and acetyl-CoA, acyl-CoA, and others are the activated forms participating in fatty acid synthesis as well as their degradation. Pantothenic acid is also a contituent of the acyl carrier protein. Additional acetyl-CoA and choline form acetylcholine, which is one of the most important neurotransmitters. Thus one can expect a rather high concentration of this vitamin in the brain (Schuberth *et al.*, 1965). In guinea pigs, a deficiency of pantothenic acid can lead to degeneration of the peripheral nerve myelin and lesions of the spinal cord (Reid and Briggs, 1954).

In human volunteers α-methylpantothenic acid, an inhibitor of the vitamin, together with a deficient diet, in 25–30 days, produced symptoms which included apathy, depression, and cardiovascular instability (Bean *et al.*, 1955). Although this vitamin is involved in the synthesis of acetylcholine, an extensively studied neurotransmitter, no direct correlative data between the deficiency of the vitamin and acetylcholine levels have been reported.

7.10. CHOLINE DEFICIENCY

Generally, choline is not included in the list of vitamins but, undoubtedly, it is an essential compound, at least in the functioning of the brain. It can be made in the body by successive methylation of ethanolamine but these reactions seem to be very inactive in the brain (Ansell and Spanner, 1967). Normally choline can be synthesized by the following pathway:

serine $\xrightarrow{CO_2}$ ethanolamine $\xrightarrow{ATP}$ phosphorylethanolamine $\xrightarrow{SAM}$ monomethylethanolamine

$\rightarrow$ dimethylethanolamine $\rightarrow$ trimethyl ethanolamine (choline)

The ethanolamine and choline are found in the form of their corresponding phospholipids, e.g., phosphatidylethanolamine and phosphatidylcholine (also called lecithin).

The importance of choline to the CNS became apparent when Cohen and Wurtman (1976) showed that dietary intake of choline for 3 to 11 days increases choline levels in the serum and the brain; and, more important, acetylcholine levels also increase in the brains of experimental rats. Many earlier observations regarding diet-related increases in the composition of the brain were true, in most cases, only during the early growth periods. But the above results were seen in adult rats when their brain growth spurt was long past. This observation, therefore, opened up a whole new field dealing with effects of nutrition on mature brain. Up until recently these and similar findings involving dietary tryptophan and brain 5HT levels (Fernstrom and Wurtman, 1971a–c), it had been assumed that the mature brain was little influenced by dietary intake. Wurtman (1979a) has listed certain prerequisites for precursor levels in plasma (and these may have to be derived from dietary intake or made in the body by nonneuronal tissue and released in the circulating blood) to affect neurotransmitter levels. The prerequisites include a capacity to influence plasma levels of precursor molecules and a system for the uptake of this precursor from the blood into the brain, preferably by a low-affinity transport system. The synthesized neurotransmitter must have available extra receptor sites for physiological action.

The free choline pool forms a ready source for the synthesis of acetylcholine in the brain since the enzymes that are involved in methylate ethanolamine methylation are not present to any great extent in the brain. Haubrich and Chippendale (1977) report that the level of free choline in the rat cerebrum is about 0.024 μmole/g fresh cerebrum, whereas that of phosphatidylcholine is 14.7 μmol/g or about 600 times more. Thus, there is a need to release choline from brain phosphatidylcholine, and, according to a report by Miller et al. (1977), brain lecithin turns over with a half-life of 3 to 4 days. The other reactions by which choline could be released into the free choline pool could be hydrolysis of sphingomyelin and the base exchange reaction; but, the contribution from these reactions to the overall supply of choline has not yet been quantitatively determined. However, the need for such a choline source is clearly indicated by the rate of synthesis of acetylcholine, ~6 nmole/min/g (Gibson et al., 1978), and the relatively slow rate of entry of plasma choline into the brain, ~0.3 nmole/min/g (Trabucchi et al., 1975). Obviously any choline that is released is reutilized not only for synthesis of acetylcholine but for storage as choline-containing phospholipids.

Sources of choline in the diet (Wurtman, 1979b) may be in the form of free choline (choline chloride), lecithin, or sphingomyelin. Butter provides about 150 mg of lecithin and 100 mg of sphingomyelin per 100 g; whole milk contains 5–6 mg of choline chloride, 50–100 mg of lecithin, and 30–70 mg of sphingomyelin per 100 g; and eggs contain only a small amount of choline chloride (0.4 mg) but are very rich in lecithin, ~400 mg. Calf liver is a rich source of free and bound choline—650 and 850 mg/100 g, respectively. Only soybeans and oatmeal come close to these values. Wurtman (1979b) has calculated that persons consuming a typical diet take in about 3.16 g of total lecithin and 0.009 g of total unbound choline per day. Obviously these values vary from person to person in relation to the many different diets consumed by a population.

The finding of a definite correlation between choline intake and brain acetylcholine levels in the rat can have a direct application in human population. In humans, a brain disease known as tardive dyskinesia is attributed to an inadequate release of acetylcholine (Growdon et al., 1977). In patients suffering from this condition, dietary supplements of choline chloride or lecithin were indeed found to be responsive (Hirsch et al., 1978; Wurtman et al., 1977). Choline chloride is rapidly degraded by the intestinal flora in the human intestine to form trimethylamine. This compound gives off a fishy odor and thus is very distasteful, but lecithin is, of course, odorless (Hirsch and Wurtman, 1978). Lecithin circulating in the blood seems to be incorporated into the brain lipids. For example, Dhopeshwarkar and Mead (1973), while studying the transport form of lipids entering the brain, found that intravenously injected [1-^{14}C]dipalmitoyllecithin was directly incorporated into brain lipids. Similar results were obtained following intraperitoneal injection (Dhopeshwarkar et al., 1973) but the magnitude of the brain uptake was reduced. Illingworth and Portman (1972) found that administration of lysolecithin to animals brings about a rapid uptake and hydrolysis of fatty acids and choline in the brain tissue. The uptake of choline into the brain by a carrier system does not seem to be saturable even in the presence of high plasma levels (Freeman et al., 1975).

Acetylcholine levels in the brain have been implicated in the physiological basis of memory. For example, it is possible to manipulate learned behavior in animals by using drugs that alter the availability of acetylcholine in the brain (Signorelli, 1976). Such drugs can be administered to humans with no lasting effects and very low risks. However, prior to receiving these drugs, they receive methscopolamine to block the peripheral cholinergic side effects. Investigations with both anticholinergics and cholinomimetics can profoundly affect storage and retrieval of information in memory (Davis *et al.*, 1978). Neuropharmacological testing has shown that lecithin can suppress choreic movements in patients with tardive dyskinesia (Growdon *et al.*, 1978). Preliminary reports on clinical trials using lecithin in Huntington's chorea and senile dementia have been encouraging (Autuono *et al.*, 1979), but large scale testing is necessary before final conclusions can be reached.

Etienne *et al.* (1979) gave pharmacologic doses of lecithin for 4 weeks to seven patients (42–81 years old) diagnosed as suffering from Alzheimer's disease. During the treatment, the plasma levels of choline increased about fivefold and three patients showed a partial selective improvement in learning ability and the improvement disappeared as the treatment was discontinued. Thus, it seems that in certain conditions that may have origin in abnormality in cholinergic neurons, lecithin treatment may have a favorable outcome.

7.11. VITAMIN A (RETINOL, RETINAL, AND RETINOIC ACID)

McCollum and Davis in 1913 found that rats failed to grow on semisynthetic diets containing lard as a source of fat; but, when egg yolk or butter was added to the diet, growth resumed. Thus, the fat-soluble factor present in eggs or butter was a necessary component of the diet. This fat-soluble factor was finally isolated by Karrer in 1931 from fish liver oils and chemically characterized as vitamin A (Fig. 26).

In animal products, vitamin A usually occurs as the alcohol and is stored in the liver (approximately 95% of the total vitamin A in the body is stored in the liver) as vitamin A palmitate. The vitamin is slowly destroyed by heat and oxidation in the presence of light. The carotenoid pigments found in plants can be converted to vitamin A in the animal body. There are several carotenoid pigments in plants but only β-carotene has two symmetrical β-ionone rings (necessary for biological ac-

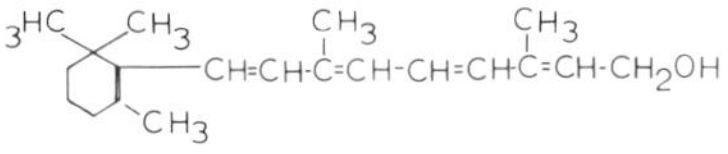

Figure 26. Vitamin A (retinol).

tion) and thus, theoretically, can give rise to two molecules of vitamin A. In the body, the conversion is not so efficient and, therefore, only about half as much activity can be expected (Moore, 1957).

7.11.1. Biological Activity and Function

It has been known for some time that in the absence of vitamin A, cessation of growth and ultimately death will occur in experimental animals. Other essential functions of vitamin A include normal vision, reproduction, growth and maintenance of differentiated epithelia, and mucus secretion (Goodman, 1980). Retinol esters are stored in the liver and, on mobilization, retinol forms a complex with retinol binding protein which in turn interacts strongly with prealbumin. This is the form in which the vitamin is transported in the plasma. Retinol binding protein is a single polypeptide (mol. wt. 20,000) with a single binding site for one molecule of retinol. Vitamin A has recently been shown to regulate biosynthesis of glycoproteins which are common constituents of cell membranes (Wolf et al., 1979). It is, therefore, easy to understand the correlation between deficiency of vitamin A and integrity of the epithelial cells.

The role of vitamin A in the visual cycle has been studied by Wald (1968). The human retina contains two types of receptor cells, rods and cones. Animals that have only rods do not have color vision. The rods contain rhodopsin, a thermolabile glycoprotein, which dissociates to form opsin and retinal when exposed to light. The retinal is an all-*trans* compound. This aldehyde is reduced to all-*trans*-retinol and then isomerized to form 11-*cis*-retinol. This *cis* compound (transported in the blood as retinol binding protein complex) is converted to the aldehyde form and combines with opsin to regenerate rhodopsin. In vitamin A deficiency both the level and regeneration of rhodopsin are decreased leading to night blindness.

In PCM, the level of retinol binding protein is low and even in the event of adequate dietary intake of vitamin A, the plasma vitamin A levels are low, indicating that release and transport function are impaired. Under these conditions, there could be a decreased level in the brain and the consequences of this have been observed by several workers in this field. Wolbach and Hegstead (1952) reported that vitamin A-deficient chicks developed hydrocephalus with increased intracranial pressure due to failure in the growth of calvaria. Howell and Thompson (1967) found increased growth of periosteal bone leading to obstruction of the CSF circulation in vitamin A deficiency. Reports of abnormality in the laying down of myelin following vitamin A deficiency have been documented (Clausen, 1969) but others could not confirm these results (Kern, 1970). Vitamin A toxicity from excess ingestion occurs due to the fact that the binding capacity of the retinol binding protein is exceeded and retinol circulates in the free form. This free form, in contrast to the bound form, seems to be toxic. Acute vitamin A toxicity may result following ingestion of 300,000 IU during infancy. This leads to nausea, vomiting, and drowsiness (Anderson and Fomon, 1974).

7.12. VITAMIN E (TOCOPHEROLS)

In 1922 Evans and Bishop discovered a fat-soluble factor that prevented resorption of fetuses in pregnant rats and testicular degeneration in male rats. It was then known as the antisterility vitamin. Later, a new word, tocopherol, was coined for this vitamin, having its genesis from the Greek words *tocos* (childbirth) and *phero* (to bear). The biological influence of the tocopherols reached far beyond reproduction as new knowledge about this vitamin accumulated through active research (Fig. 27). The tocopherols are characterized by substitutions on the ring, as shown in Table 5.

The tocopherols show UV absorption which can be used in analytical methods. Other methods include the colorimetric method using O-phenanthroline or α,α'-dipyridyl; but high-performance liquid chromatography has replaced older methods due to very high sensitivity and high resolution capacity.

7.12.1. Biological Function

α-Tocopherol is considered a highly active physiological antioxidant that prevents oxidation of polyunsaturated lipids, but another role of this vitamin is interrelated with the metabolism of selenium. In the first role, vitamin E would protect the cellular and subcellular membrane phospholipids since these lipids usually have a relatively high amount of PUFA at the 2 position. Selenium, on the other hand, is an integral part of the enzyme glutathione peroxidase. This enzyme protects the cell membrane by destroying hydrogen peroxide and, possibly, peroxides

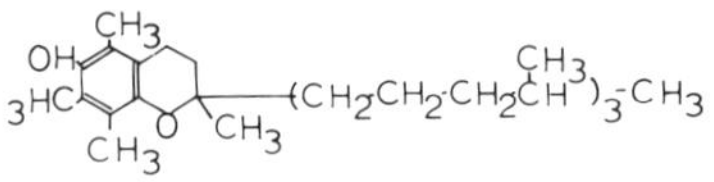

Figure 27. Vitamin E (α-tocopherol).

Table 5. Structures and Sources of Tocopherols and Related Substances

Name	Source	Groups		
		R_1	R_2	R_3
α-Tocopherol	Wheat germ	CH_3	CH_3	CH_3
β-Tocopherol	Wheat germ	CH_3	H	CH_3
γ-Tocopherol	Corn	H	CH_3	CH_3
δ-Tocopherol	Soybean	H	H	CH_3
7-Methyltocol	Rice	H	CH_3	H
5,7-Dimethyltocol	Rice	CH_3	H	H

that are formed from the PUFA. Reduction of lipid peroxides to nontoxic hydroxy fatty acids not only protects membranes but also prevents the decomposition of these peroxides into free radicals that can propagate the peroxidation chain (Tappel, 1974). With these broad functions in mind, one could hypothesize that, under adverse conditions, such as exposure to oxidative atmospheric pollutants, there would be a greater need of an antioxidant such as vitamin E to prevent peroxidation of membrane lipids. From this point of view, one can predict a relatively strong affinity between vitamin E and phospholipids of mitochondrial and microsomal membranes. The vitamin appears to be present at these sites to prevent initiation and propagation of lipid peroxidation (Scott, 1980).

7.12.2. Vitamin E Deficiency (Animal Studies)

Einarson and Telford (1960) reported that vitamin E deficiency in rats, mice, pigs, rabbits, and monkeys causes many CNS-related changes, e.g., demyelination, increase of glial cells, and loss of the Nissl substance. Verma and King (1967) found that vitamin E deficiency during gestation produced abnormalities in the embryos (halfway through normal gestation and at term) such as hydrocephalus, reduction in neurons, and abnormal choroid plexus. They also suggest a breakdown of the blood–brain barrier. Progressive muscular dystrophy has been noted by many investigators in dietary-induced vitamin E deficiency (Bunyan *et al.*, 1967). However, muscular dystrophy produced in animals by dietary lack of vitamin E, is not the same as muscular dystrophy in humans. The former responds completely with a reversal of the disease following ingestion of vitamin E but there is no effect at all on the human disease (Scot, 1980). In response to a hypothesis that the requirement for vitamin E needs to be increased during increased dietary intake of polyunsaturated fat, Witting (1974) has shown that the relationship of vitamin E and PUFA in the diet and tissue is complex but the higher requirement is indeed proved to be true.

In our own studies (Dhopeshwarkar *et al.*, 1981), it was found that in the presence of a limited amount of EFA, a lack of vitamin E in the diet over a prolonged period will not only result in unmeasurable levels of vitamin E in plasma and tissues but will lead to an accumulation of eicosatrienoic acid ($20:3\omega9$) in the liver and plasma. This suggested that in the absence of vitamin E, EFA deficiency is rapidly produced. It is not clear yet whether such a condition will develop in the presence of higher levels of linoleate in the diet.

It has been suggested that an optimal ratio of vitamin E/PUFA in the diet be maintained but there is no agreement on the suggested value of this ratio (Thompson *et al.*, 1973; Hashim and Asfour, 1968). Jäger (1975) has concluded that, under normal conditions, when the membranes are adequately "saturated" with EFA, except at very high intakes, dietary linoleic acid has little effect on vitamin E requirement. Thus, he suggests no fixed vitamin E/polyene ratio needs to be maintained in the everyday diet.

7.12.3. Relationship of Selenium and Vitamin E

Schwarz and Foltz (1957) found that selenium as well as vitamin E could prevent necrotic liver degeneration produced in rats by using vitamin E-deficient diets. Thus, diseases like nutritional muscular dystrophy in ruminants and exudative diathesis in chicks seem to respond to selenium as well as to vitamin E. However, classic experimental diseases such as muscular dystrophy in the rabbit and fetus resorption in rats could not be cured by selenium alone. The controversial situation was finally cleared by Thompson and Scott (1969) who found that if the diet was exceptionally low in selenium (<0.005 ppm), vitamin E could not replace the element in curing a deficiency disease such as pancreatic fibrosis which was brought about solely from selenium deficiency. This newly discovered disease is characterized by poor growth, loss of hair, and reproductive failure and responds specifically to dietary selenium (McCoy and Weswig, 1969). This newly acquired knowledge clarified the nutritional requirement of selenium independent of vitamin E.

As mentioned earlier, glutathione peroxidase (EC 1.11.1.9) is an enzyme that contains selenium as an integral part. This enzyme catalyzes the reduction of certain hydroperoxides of PUFA to the corresponding hydroxy compounds. It has a molecular weight of about 88,000 and contains four atoms of selenium per mole of protein. It has been reported that the primary function of this enzyme is reduction of hydroperoxides of unsaturated fatty acids, thus preventing the propagation of the free radical chain reactions at least in the capillary membranes (Noguchi *et al.*, 1973). Some doubt has recently been cast on this function and a soluble, heat-labile, glutathione-dependent factor has been proposed for the reduction of PUFA hydroperoxides (McCay *et al.*, 1981).

7.12.4. Human Studies

7.12.4.1. *Vitamin E Deficiency and Retrolental Fibroplasia*

Retrolental fibroplasia is considered to be a disease of the growing blood vessels. In humans the growth of the blood vessels in the retina occurs *in utero* (Michaelson, 1948). Therefore, premature infants are more prone to have this disease than full-term infants. By the fourth month of gestation, the retina thickens and diffusion of oxygen and nutrients no longer can satisfy all needs, so there is a need to develop and increase capillary blood supply. The severity of the disease is correlated with a degree of prematurity and is increased by hyperoxia (Johnson *et al.*, 1974). Presently, it is advised not to exceed ambient oxygen tension about 40% to eliminate this danger from high oxygen tension. Clinical trials with vitamin E suggest that the vitamin reduces the overall incidence and decreases the severity of acute disease in premature infants. The greatest benefit occurred in premature infants weighing less than 1500 g at birth (Johnson *et al.*, 1974).

7.12.4.2. *Batten's Disease*

Mental retardation and seizures are common symptoms of Batten's disease and examination of brain composition indicates an accumulation of a lipopigment, ceroid. This seems to be a slightly different pigment from lipofuscin, the so-called "age" pigment. It is thought that this pigment arises from massive peroxidation of EFA since α-tocopherol is an excellent biological antioxidant, trials were conducted to determine the effect of vitamin E supplementation in such patients (Zeman *et al.*, 1970). The standard therapy includes administration of a mixture of vitamin E, ascorbic acid, methionine, butylated hydroxy toluene and has been found to have some beneficial value. However, Siakotos *et al.* (1974) found that blood and tissue levels of α-tocopherols in patients were significantly higher than age-matched controls. Thus, these authors question the relevance of vitamin E supplementation. They suggest that vitamin E may have biochemical role in addition to just being an antioxidant to prevent peroxidation of lipids.

7.12.5.   Recent Studies on Auto-oxidation from Atmospheric Pollutants

Until recently hydroperoxides were considered as products of auto-oxidation of PUFA; this certainly is true when one is dealing with bulk phase. However, recent work by Wu, Stein, and Mead reviewed by Mead (1980) showed that in monolayer systems, which are closer to membrane bilayers in cells, epoxides were formed rather than hydroperoxides. Such epoxides were formed from any olefinic compound including oleic acid and cholesterol. (The α-5,6-cholesterol epoxide has been shown to be carcinogenic.)

The epoxidation was retarded by α-tocopherol but when concentration was reduced to about 0.004 mole% the oxidation of linoleate continued at a rate equal to that when no α-tocopherol was present. Calculation of the protection offered by α-tocopherol indicated that 1 α-tocopherol molecule protected about 20,000 molecules of unsaturated fatty acids. In studying peroxidation of microsomal membranes, Mead and co-workers (1980) found that, in microsomes, fatty acids were protected if the animal had received an α-tocopherol-supplemented diet. The system used consisted of microsomes incubated in the presence of iron–ascorbate (a normal initiator of tissue peroxidation) or 12 ppm of NO_2. Under these conditions, NO_2 accelerated the destruction of α-tocopherol but not the PUFA. The effects of atmospheric pollutants on membrane fatty acids in the lung and the RBC has been studied by many workers but these cells are not rich in PUFA as compared to the cells in the brain. However, very little work using brain-cell culture systems has been done with regard to effects of atmospheric pollutants.

7.13.   VITAMIN D

A fat-soluble factor that was necessary for growth and prevention of xerophthalmia was named fat-soluble A and another dietary fat-soluble factor that cured rickets and was relatively resistant to oxidation was called fat-soluble D. Later work showed that this factor was produced in the skin by exposure to UV light. Identification and characterization of this factor led to its present chemical form now known as vitamin D_3.

7-Dehydrocholesterol $\xrightarrow[\text{irradiation}]{\text{UV}}$ previtamin $D_3 \longrightarrow$ vitamin D_3
(present in the skin)

Vitamin D is transported protein bound in plasma as vitamin D_3. The vitamin is absorbed along with dietary fat and accumulates in the liver and is converted to 25-hydroxyvitamin D_3. This reaction is catalyzed by cytochrome P-450-dependent mixed-function monooxygenase (Madhok and DeLuca, 1979). A second reaction, again a hydroxylation, converts 25-hydroxycholecalciferol to the 1α,25-dihydroxy compound. But this reaction occurs mainly in the kidney mitochondria (DeLuca and Schnoes, 1976). The 1,25-dihydroxyvitamin D_3 is the biologically active form of vitamin D_3. Biological function of this vitamin is related to elevation of intestinal calcium and phosphorus transport in conjunction with parathyroid hormone and a mobilization of calcium and phosphorus from the bone.

Frank deficiency of vitamin D_3 is rare in human populations with access to a normal variety of foods. In most countries fortification of milk and bread is common and exposure to sunlight can compensate for minor deficiencies. Recently a cytosolic receptor specific for 1,25-dihydroxyvitamin D_3 has been found in the pituitary gland of rats (Anonymous, 1981). In view of this observation, researchers will soon be looking for the role of vitamin D_3 and its metabolites in neurophysiology and biochemistry.

MINERAL DEFICIENCIES AND EXCESSES

8.1. ZINC

Zinc was first shown to be essential for growth of living organisms in 1869. In 1934, it was shown to be essential for growth and normal development of mammals (Todd *et al.*, 1934). In 1940, the first metalloenzyme containing zinc was isolated (Keilin and Mann, 1940) and since then several other enzymes have been shown to contain zinc (Parisi and Vallee, 1969). Recently, it has been reported that zinc and iron are nutritionally interrelated. The highest concentration of zinc in the body was found in layers of the choroid in the eye. This layer contains an enzyme retinene reductase which is a zinc metalloenzyme.

Zinc deficiency can be brought about by the use of plant-derived foods that contain phytic acid. Phytates form insoluble complexes with zinc and casein, which are resistant to digestion by proteolytic enzymes (O'Dell and Savage, 1960). In rats, excess dietary zinc caused a marked loss of iron resulting in anemia (Cox and Harris, 1960). In humans, use of galvanized iron utensils may bring about similar excesses of zinc. This is reported to produce lethargy and loss of muscular coordination. Zinc deficiency in rats resulted in growth retardation, lymphocytopenia, testicular atrophy, and bone deformities. Hurley (1969) was first to note that zinc deficiency caused brain malformations. This was confirmed by Sandstead and co-workers (1971), who reported that brain DNA, protein, and lipid concentrations were significantly decreased in pups born to dams on a zinc-deficient diet. Zinc seems to be an essential nutrient for the synthesis of nucleic acids and proteins. Turhune and Sandstead (1972) found that in zinc-deficient pups the activity of DNA-dependent RNA polymerase decreased after the tenth day of life. The decrease was further shown to be dependent on zinc deficiency rather than on partial starvation or decreased food consumption.

Zinc deficiency produced in pregnant rats has a profound effect on fetal development: generally malformation involves almost every organ. More alarming is the fact that even if the deficiency occurs during the middle of the gestation period (6–14 days) fetuses are small and malformed (Winick, 1976).

Prasad et al. (1961, 1963) have reported effects of zinc from a study of adolescent males from the Nile delta of Egypt. The diet of these individuals was predominantly wheat, which is high in phytate. By forming complexes with iron and zinc, the phytates prevented absorption from the gut. Geophasia, a practice of mixing clay with food, has been noticed in dwarfed individuals. The clay may combine with zinc and iron by a cation-exchange mechanism (Halstead and Smith, 1969). High temperature may also be a factor in causing zinc deficiency since sweat contains rather high quantities of zinc (1 mg/liter). Primary features of the deficiency are growth retardation and hypogonadism, and, in a few cases, hypopituitarism was indicated (Sandstead et al., 1967). Similar results were reported in a study in Iran confined to females (Halstead et al., 1972) and later to school boys (Ronaghy et al., 1969). In the United States, the deficiency of zinc, due mainly to marginal dietary intake, has been reported in a study from Denver (Hambidge et al., 1972).

Zinc deficiency has been shown to cause behavioral changes in animals. Rats raised on a zinc-deficient diet showed a decreased tolerance to stress and female rats seem to be more aggressive (Halas and Sandstead, 1975; Halas et al., 1975).

8.2. COPPER

Both copper and iron are required in mammalian nutrition to prevent anemia. Similar to zinc, copper also forms metalloproteins and enzymes from which copper cannot be dissociated without loss of activity. Some notable cuproenzymes are cytochrome c oxidase, dopamine β-hydroxylase, superoxide dismutase, and tyrosine oxidase. Copper seems to be essential for iron absorption and hence can have an effect on hemoglobin synthesis. It has been proposed that a copper-dependent enzyme, ferroxidase, converts Fe^{2+} to Fe^{3+}, which is the storage form of iron. Copper deficiency is known to produce connective tissue dysfunction, cardiovascular disorders, and bone fragility (Underwood, 1971). The defect in bone formation is traced to decreased levels of lysyloxidase which is involved in crosslinking to form collagen. More serious damage to the cardiovascular system has been noted in copper deficiency such as massive internal hemorrhage from the rupture of large blood vessels or even the heart (O'Dell, 1976,a,b).

Copper deficiency in soil and grass produces ataxia in lambs, "swayback" in cattle, and posterior paralysis in swine. The underlying cause seems to be a lack of myelination of the nerves. DiPaolo and Newberne (1974) found that in copper deficiency myelin lipids, sulfatides, and cerebrosides were reduced significantly. However, direct involvement of the neurons is also possible since concentration of neurotransmitters is low in copper deficiency. The probable cause of this is related to the fact that catecholamine β-hydroxylase is a copper metalloenzyme (O'Dell, 1976a).

Menkes and co-workers (1962) described a childhood syndrome inherited as a sex-linked trait, characterized by mental retardation and seizures, and was usually fatal within a span of about 3 years. Abnormally low levels of serum and tissue copper levels were found. But oral administration of copper was not successful in clinical treatment of such patients. Other instances of copper-deficient infants have been recently described by Grover and Scrutton (1975).

Wilson's disease (hepatolenticular degeneration disease) is a familial inherited disorder. The frequency is estimated to be 1 in 500. In this disease, copper accumulates in a number of organs, including the brain. CNS involvement is suggested by symptoms such as progressive rigidity, intension tremor, and corneal degeneration. Although copper accumulates in tissues, the serum copper levels are decreased. Since there is an accumulation of copper in tissues, promotion of copper excretion and control of dietary intake would be a rational approach of treatment. Therapeutic use of penicillamine has been reported to bring about marked clinical improvement, particularly of the neurological manifestations. Patients who were asymptomatic to begin with do not develop symptoms if treated prophylactically (Goodman and Gilman, 1975). Recently, cell culture studies have revealed a possible diagnostic marker. Skin fibroblasts from patients with Wilson's disease show an elevated copper concentration and a decreased ratio of copper to protein, which has a molecular weight equal to or greater than 30,000 (Chan et al., 1980).

So far, emphasis has been put on mineral deficiency and its effect on general metabolism, with special reference to deleterious effects on the nervous system wherever applicable. We now turn to serious toxic effects by contamination with excess minerals, with special attention to mercury and lead which have been the center of recent research.

8.3. MERCURY

Various forms of mercury are used in agriculture, industry, and even pharmacology. A recent increase of interest is toxicological effects is due to outbreaks of mercury poisoning, involving large numbers of people.

Mercury occurs in the form of free metal, monovalent mercury salts, and divalent organic derivatives. Mercury can be toxic in all these forms. For example, the free metal, although it boils at extremely high temperature (357°C), vaporizes very easily at room temperature (25°C). One cubic meter of air saturated with mercury vapor contains 19.5 mg of Hg at room temperature (25°C). Mercurous chloride (HgCl—also known as calomel) is the most common monovalent salt. On exposure to light or bacterial action, it is readily oxidized to mercuric chloride ($HgCl_2$), also called corrosive sublimate. Organic forms such as ethylmercury, C_2H_5—HgCl, are used as a seed sterilizer (seed dressing) and dimethyl mercuric sulfide, CH_3—S—Hg—CH_3, and aryl mercury, C_6H_5COOHg—$OOCH_5C_6$, are other known forms commonly used. In pharmacology, mercurial diuretics have been known for several years and have a general structural formula,

$$R—CH_2—CH—CH_2—Hg—X,$$
$$|$$
$$O—Y$$

where R, Y, and X could be different substituents. Mersalyl and chlormerodrin are well-known diuretic compounds. In industry, mercury is used in various instruments and in UV light sources. In short, this highly toxic element is commonly used in many ways in modern science and technology.

8.3.1. Absorption of Mercury

1. *Pulmonary:* Mercury vapor easily penetrates the alveolar membrane and thus enters the blood in the lungs. Some small amounts of organic mercurial compounds, including methyl mercury chloride, vaporize quickly and can be absorbed easily by inhalation.

2. *Gastrointestinal Absorption:* Only about 7% of ingested inorganic salts are absorbed by humans compared to about 95% of the administered dose of methyl mercury (Miettinen, 1973). However, unlike lead, administration of methyl mercury along with dietary fat does not influence its absorption in the gut.

3. *Percutaneous Absorption:* Methyl mercury compounds are also absorbed via the skin and can cause serious harmful effects.

Dangerous harmful effects of mercury to fetus and newborn can be explained by the fact that the transport of short-chain alkyl mercury is not restricted by the placental barrier. Methyl mercury is more easily transferable than inorganic mercury in the rat (Mansour et al., 1973). Reynolds and Pitkin (1975) have shown that in rhesus monkeys, the transport of methyl mercury from the mother to the fetus occurs more readily than in the reverse direction. Thus, there is a slow buildup of mercury in the fetus. This placental transfer phenomenon in the monkey is very similar to that in human conditions. Further, the transported methyl mercury seems to accumulate in the fetal brain (Wannag, 1976).

Since the toxicity of various forms of mercury is different, one has to know if various forms are interconvertible. For instance, alkyl mercurials are highly toxic, aryl compounds, in particular; phenylmercury compounds are much less toxic; and organomercurials (e.g., diuretics) are less toxic still and, in fact, are used as drugs. Elemental mercury ($Hg°$) is oxidized to mercuric ion (Hg^{2+}) in the blood as well as in tissues, probably via the action of the enzyme catalase. The reverse reaction is caused by certain common micro-organisms (Magos et al., 1964). The C-Hg covalent bond in methyl mercury is rather resistant to cleavage but that of phenylmercury or ethylmercury is readily cleaved by rat liver and kidney (Fang, 1974), and this activity was shown to be increased by dietary selenite. In large lakes, even mercuric chloride can be converted to methyl mercury by the microbial flora. Thus, it seems that, in general, all forms of mercury eventually can form highly toxic materials, some rather quickly, others after a long period.

Once incorporated into body tissues, inorganic mercury and organomercurial compounds are excreted in the bile and partly in the feces. Some of the mercury is reabsorbed in the enterohepatic circulation of the bile but this reabsorption is decreased by oral administration of polythiol resin (Clarkson et al., 1973). Administration of 2,3-dimercaptopropanol (BAL) also helps to increase biliary excretion of methyl mercury. Mercaptodextran (mol. w. $\sim$10,000) helps to eliminate mercury from the body including the brain (Aaseth, 1973). Another mode of excretion of mercury is via the kidney and urine. Urinary excretion in a patient exposed to mercury vapor was remarkably elevated by D-penicillamine (Suzuki and Yoshino, 1969).

Large-scale outbreaks of mercury poisoning have been reported from Japan, Iraq, Pakistan, and Sweden. The first of these outbreaks occurred in Japan, the epidemics of Minamata Bay and Nigata. A total of 121 people living in villages around Minamata Bay suffered toxic effects during 1953–1960. Twenty-two infants were poisoned prenatally and 46 people died. In Nigata, similar outbreaks with 6 deaths were reported. Investigation of the cause of this large-scale outbreak revealed that the fish, which formed the basic food of these people, was contaminated with methyl mercury. The methyl mercury in the waters of Minamata Bay came from the dumping of waste water from a plastic factory that was using mercury compounds as catalysts. Since then, the disease due to mercury poisoning bears the name *Minamata disease.*

Discharge of phenylmercury compounds from the wood pulp industry and metallic and inorganic mercury from chlorine-alkali plants, where mercury electrodes are used, have all contributed to contamination of waters in lakes and rivers in many countries. This has led to increased amounts of mercury in fish as well as in game birds. Further, the mercury in these sources was found entirely to be methyl mercury, the most toxic form of this element. Subsequent research by many investigators clearly established that microorganisms present in the sediments of lakes and rivers were capable of transforming various forms of mercury into methyl mercury. Ackefors (1971) has pointed out that in Sweden there was a high concentration of mercury in the fish, so much so, that fish from certain areas were forbidden to be part of the human diet. The mercury contamination of food was widespread as seen from the following typical values.

Swedish eggs	0.029	Swedish pork chops	0.03
Norwegian eggs	0.020	Swedish pig liver	0.06
Danish eggs	0.004	Swedish beef	0.012

These values were obtained before restrictions on the use of mercury compounds. As mentioned before, Hg^{2+} was converted to $(CH_3)Hg^+$ and further to $CH_3\text{-}Hg\text{-}CH_3$; dimethyl mercury is volatile and lipophilic, adding to the toxicity.

The effects of methyl mercury poisoning can be described as loss of coordination and difficulties in eating, hearing, speaking, and vision. Animal experiments show that mercuric chloride can have a toxic effect on the blood–brain barrier (Steinwall, 1969), whether such an effect is possible in humans is not known.

Another large-scale epidemic of mercury poisoning occurred in farming families in Iraq (Bakir *et al.*, 1973). A total of 6530 cases of poisoning were recognized and 459 hospital deaths occurred, all attributed to methyl mercury poisoning. Mercury-treated wheat and barley were imported into Iraq and used to make bread. Some of it was fed to farm animals and part of it got mixed with the soil. The half-life of mercury in patients varied from 65 to 105 days. D-Penicillamine, BAL, and other compounds were tried as binding agents. In some individuals there was a dramatic reduction in blood levels but in others it was a very slow process. A thiol resin given at 8 g/day was found to be nontoxic and did not redistribute the mercury in the body; in most cases it helped in the excretion process.

Mercury poisoning may not show any serious specific signs in adults. For example, mothers themselves may not show toxicity (Amin-Zaki *et al.*, 1974) even though the mercury level in the tissue and blood may be elevated; but the offspring almost always shows signs of toxicity. The blood, urine, and hair levels in infants born to mothers exposed to mercury are higher (Bakir *et al.*, 1973). The overall syndrome in infants is characterized by severe mental and motor retardation, irritability progressing to myoclonic seizures, loss of sight and hearing, and abnormal posture (Amin-Zaki *et al.*, 1974). Biochemical abnormalities include inhibition of carbohydrate oxidation (Menon and Kark, 1976). Doses of 40 mg/kg of methyl mercury on the 14th day of pregnancy resulted in stillbirths in rats but 10 mg/day did not cause any gross abnormalities (Kark *et al.*, 1977). The incorporation of β-hydroxybutyrate into lipids was significantly decreased. Other biochemical parameters have not been fully examined to determine the underlying cause of neurotoxicity following methyl mercury exposure.

Mercury poisoning can be avoided rather easily now that large outbreaks have occurred and its toxicology known. However, simple methods, like coloring the "dressed" seeds with a dye, may not be enough. The farmers may wash off the dye but not all of the mercurial compounds. Hazard signs clearly marked and warning labels in the language of the country are an absolute must. The epidemics bring forth the necessity of close cooperation between industry, agriculture, and large populations, including the farmers, to avoid any accidental exposure to this highly toxic material.

8.4. LEAD

Toxicity of lead (Pb) to human populations is not a newly discovered finding. It has been known since early Egyptian and Roman civilizations. Thomas and Blackfan (1914) first described the pathological effects of lead-induced encephalopathy in children and thus started the reinvestigation of lead toxicity in terms of modern medicine and biochemistry.

Measuring lead content of the hair has been used to determine exposure to lead toxicity. Recently Weiss *et al.* (1972) have reported that in spite of the introduction of tetraethyl lead in gasoline in 1923, which would tend to increase exposure to lead, the data on the lead content of the hair do not support increased environmental hazard. They reported that "antique" hair (years 1871–1923) contained a significantly higher content of lead than did rural and urban population hair in 1971. However, since exposure to organic lead (tetraethyl lead additive) is more harmful than inorganic lead, the present trend is to decrease or eliminate tetraethyl lead as an antiknock additive and replace it with non-lead-containing compounds.

The ingestion of lead is closely related to pica, a child's habit of chewing on sides of cribs (then made from metal), furniture, and woodwork and eating painted plaster and fallen paint flakes. The blood levels of lead in 5-year-old children with this habit can be greater than 50 μg/dl. It is possible that absorption of lead is higher in children than in adults, and thus it becomes another contributory factor (Alexander, 1974). In addition to ingestion of leaded paint, improperly glazed ceramics may contribute, although not quite so commonly, to exposure to lead. An increase in lead in the soil or street dust may be the result of exposure to automotive exhaust (Lepow *et al.*, 1974). It should be mentioned here that particle size has an important bearing on lead toxicity. For instance, fine dust on the interior of the house may be more dangerous than larger particles in the soil.

The scheme in Figure 28 assumes that "diffusible plasma lead" is of sufficiently small molecular dimension to permit transport across the cell membranes. The potential toxic effects of lead exposure come from its deposition in soft tissues such as liver, brain, and kidney rather than hard tissues like bone or teeth. Exposure to organic lead, tetraethyl lead,

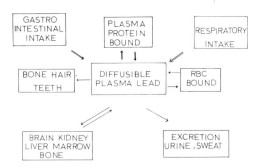

Figure 28. Metabolic fate of lead.

can add to respiratory intake and as mentioned before may be more toxic than inorganic lead (Bolanowska, 1968).

8.4.1. Pathological Effects

Hematological: The clearest indication of lead poisoning comes from the anemic condition after exposure to toxic amounts of lead. The RBC are microcytic and hypochromic (as in iron deficiency) even if the blood levels of iron are normal. The main cause of this is the impairment of heme synthesis and shortened life span of the RBC. δ-Aminolevulinic acid dehydratase is the most sensitive enzyme in heme synthesis pathway that is affected by lead. Another specific blockage is related to the enzyme ferrochelatase that catalyzes the incorporation of Fe^{2+} into the porphyrin ring (Órtzonsek, 1967).

Neurological: The effects of lead toxicity in the CNS result in ataxia, stupor, coma, and convulsions depending upon the exposure level. Cerebral edema and an increase in CSF pressure are also commonly encountered. In adults, the CNS involvement may not be obvious but they still show muscle weakness and wrist or foot drop (Schlaepfer, 1969). Animal studies show that lactating rats fed lead cause neurological damage in suckling pups along with the retarded growth. This is explained by the fact that lead is transmitted to the young via milk (Krista and Momellovic, 1974). However, if the lactating dams were kept on rat chow and the pups were given 1 mg of lead orally every day and weaned to a diet containing 40 ppm lead, the growth rate was no different from controls. At 5–6 weeks of age the pups receiving lead were hyperactive, and the endogenous brain dopamine levels were unchanged whereas NA levels were increased. This suggested a correlation between NA levels and increased motor activity (Silvergeld and Goldberg, 1975).

Brown (1975) administered lead acetate to dams by gavage (35 mg/kg/day). Two groups of pups were studied. One group received lead via mother's milk during Days 1–10 or Days 11–21. The group that received lead at the earlier period showed significantly lowered learning capacity compared to pups who received the dose during the later period of Days 11–21. The blood levels in the first group were almost twice as high as those in pups that received lead during the 11–21 days. The learning deficiency persisted up to 8–10 weeks, even though the lead exposure was halted at Day 10 and the blood levels returned to normal. The second group needed an almost fourfold higher dose of lead to have an effect on learning ability. Rosenblum and Johnson (1968) fed 4% lead into the diet of mice at parturition. There was a high mortality rate, growth retardation, and poorly developed righting reflex in the pups.

Michaelson and Sauerhoff (1974) were able to produce behavioral changes such as hyperactivity, aggressiveness, and repetitive grooming behavior following exposure to lead.

How much lead ingestion is needed to produce adverse effects is difficult to answer because of individual variability. Absorption of dietary lead depends, however, on the vehicle with which the lead is mixed. For example, if lead is administered in oils, fats, or milk rather than in a dry diet, the absorption of lead is significantly increased (Recommendation for Prevention of Pb Poisoning in Children, 1976). Blood levels of lead are considered good indices to monitor toxic effects. The first metabolic evidence of the toxic effects of lead (hematopoietic) appear at approximately 25–30 μg Pb/dl; however, frank anemia appears only when levels reach 50–60 μg/dl. Assuming that 50% of the ingested dose is absorbed by children 3–5 years old, a daily intake of 9 μg Pb/kg/day would be sufficient to produce a blood level of 20 μg Pb/dl (Barltrop and Killala, 1967). Paints are a source of lead, and, although they may contain only about 0.5% lead (5 mg/g which is the legal limit in the United States), using a factor of 17% absorption in the case of leaded paint, calculations reveal that children may ingest 0.7 to 2.1 mg Pb/day assuming 1–3 g of paint is ingested per week. This turns out to be 9.7–29.1 μg of lead/kg body wt. It has been found that approximately 4–5 μg Pb/kg/day gives a relatively harmless blood level of about 20 μg or less Pb/dl blood. Thus, the above example shows that the ingestions of such amounts would be far in excess of a tolerable dose. However, the legal limit of 0.5% is seldom reached in household paints sold on the market today. Only about one-tenth (or less) of this amount is found in paints sold recently. Older house paints would still pose a health hazard but this will be progressively reduced as these older houses are torn down (Recommendation for Prevention of Pb Poisoning in Children, 1976).

8.5. ALUMINUM

Aluminum is not only a commonly found element in mineral deposits all around the world, but is used in so many products that exposure to it is inevitable. Although no definite biological function has been attributed to aluminum, it is possible that aluminum might be toxic under conditions of excessive exposure. For example, McLaughlin *et al.* (1962) have described a patient working in an aluminum ball mill who developed progressive encephalopathy, dementia, and convulsions. On autopsy, the brain was found to contain 20 times the normal concentration of aluminum.

Aluminum hydroxide applied to cerebral tissue induced epileptic seizures in animals (Kopeloff *et al.*, 1942). A trace amount of aluminum injected into the subarachnoid space resulted in progressive encephalopathy and cellular changes that were very similar to those found in Alzheimer's disease occurring in man. Aluminum concentrations approaching those used in animal experimental models have been found by Crapper *et al.* (1973) in certain areas of the brains of patients with clinically diagnosed Alzheimer's disease. Recently, the aluminum toxicity and its relation to Alzheimer's disease was reiterated by findings of Perl and Brody (1980), who noted foci of aluminum within the nuclear region of a high percentage of neurons containing neurofibrillary tangles, suggesting association of aluminum to Alzheimer's disease. The authors point out that other etiological factors need not be totally excluded. Consider the report by Nikaido *et al.* (1973); these authors point out that in Alzheimer's disease, senile plaques in the brain show a selective localization of yet another metal, silicon, the most abundant element in the earth's crust. Thus, further research is needed before aluminum toxicity or for that matter, metal toxicity in general, may be established as the etiological factor in human Alzheimer's disease.

8.6. CADMIUM

Occurrence of excessive amounts of cadmium is restricted to a few industrial sites and their immediate vicinity. The only exception to this is the use of metal-contaminated digested sewage sludge and rock salt, containing higher amounts of cadmium, as fertilizers in agriculture. Large steel mills with open-hearth furnaces, zinc smelters, and electroplating plants are sources of cadmium exposure and industrial workers of these areas are exposed to cadmium for prolonged periods (Yost, 1979). Friberg (1948) was responsible for focusing attention on cadmium toxicity in industrial workers. He found renal damage and emphysema common in these workers. There was an increased excretion of glucose, proteins, amino acids, calcium, and phosphorus (Lauria *et al.*, 1972).

In 1969, a unique disease was described by Tsuchiya (1969) in Japan which he called *Itai–Itai Byo* (ouch–ouch disease). The patients complained of severe pain in the bones and had several fractures and severe osteomalacia. The cause of the disease was related to exposure to cadmium from a nearby mine factory. The factory was discharging cadmium-laden wastewater into the river, the source of drinking water for the nearby population (*Research News*, 1973). Cadmium exposure has been implicated as a cause of hypertension (Schroeder, 1965) and cardiovascular disease (Carroll, 1966).

Gabbiani *et al.* (1967) found that administration of cadmium to newborn rats caused hemorrhagic lesions in the brain, whereas in older animals, the lesions were produced in spinal sensory ganglia. Cadmium seems to cross the blood–brain barrier with relative ease in newborn rats (Lucis *et al.*, 1972). Once cadmium was introduced into the body, it seemed to accumulate in the brain as the animal aged (Stowe *et al.*, 1972). Rozear *et al.* (1971) and Cearley and Colman (1974) have reported altered behavior pattern and neural function following cadmium ingestion. This observation led to speculation about the effects of cadmium on neurotransmitter metabolism. Rastogi *et al.* (1977) found an increase in NA and dopamine in the hypothalamus, pons, and medulla, whereas in the midbrain there was an increase in NA and 5-hydroxyindoleacetic acid (5-HIAA) following exposure to cadmium.

Interestingly, the manifestations of cadmium toxicity can be overcome if selenium or zinc are given either simultaneously or prior to cadmium exposure. Thus, there may be a competitive antagonism between essential elements and those that are toxic (Singhal and Merali, 1979).

NUTRITION AND NEUROTRANSMITTERS

9.1. INTRODUCTION

In 1921, Otto Loewi was the first to show that transmission of nerve impulses was mediated by chemicals. In his classic experiments, he demonstrated that when ventricular fluid of a stimulated heart was transferred to a nonstimulated frog heart, the effects of stimulation were observed in this heart. This meant that the nerve stimulus of the first heart was reproduced by the chemical activity of the solution transferred to the second heart (Loewi, 1960). Subsequent analysis in his laboratory demonstrated that this chemical substance was acetylcholine, a rather simple and relatively small molecule. Ever since this discovery, acetylcholine has been identified as a signal transmitter that has many functions in the human body. For example, it can slow down the heart rate, constrict involuntary muscle, and participate in complex integrating functions of the brain and spinal cord. A neurotransmitter is now defined as a chemical substance discharged from a nerve fiber ending, which reaches and is immediately recognized by a receptor on the surface of a postsynaptic nerve cell, and the net effect is either a stimulation or inhibition of a receptor cell. A search for other neurotransmitter substances, for example, a substance that would increase heart rate, led to the isolation of noradrenaline or norepinephrine (NA) by von Euler (1956). The established criteria used to define a substance as a neuro-

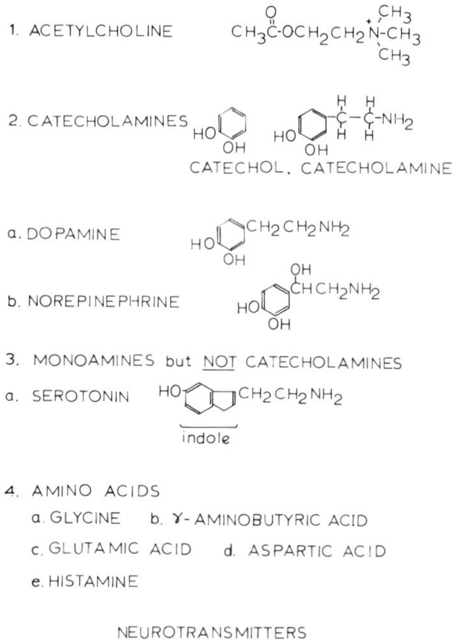

Figure 29. List of known and potential neurotransmitter substances.

transmitter include (1) secretion following stimulation of the nerve; (2) binding with a specific receptor on the postjunctional cell; (3) production of a biological response, stimulation, or inhibition; and finally (4) a mechanism to rapidly terminate the biological effect. A number of neurotransmitters that would satisfy most of these requirements have been identified (Krnjevic, 1974; Bachelard, 1974) (Fig. 29).

9.2. SIGNAL TRANSMISSION

At any given nerve muscle synapse, the gap ($\sim$500 Å) between the membranes of the nerve terminal and the muscle cell is filled with a fluid. Along the nerve terminal there are specialized areas filled with clusters of tiny vesicles (sacs) each containing about 5000 molecules of acetylcholine. In the opposing muscle cell membranes there are deep invaginations termed *junctional folds*. The ends of these folds contain receptors, specific to a given transmitter, e.g., acetylcholine. Some of the receptors also respond by binding with muscarine, a mushroom alkaloid. This process can be blocked by atropine (Bachelard, 1974). These responses are typically slow. Other receptors respond to nicotine and are blocked by drugs of the curare family.

An impulse arriving at the presynaptic nerve terminal causes mobilization of Ca^{2+} ions (even from extracellular medium) in the membrane. This leads to fusion of hundreds of vesicles with the active zone of the presynaptic membrane and simultaneous liberation of neurotransmitter substance into the synapse. The transmitter substance then binds tightly to the specific receptor on the muscle cell membrane. Within 0.3 msec, about 2000 channels are opened in the muscle cell membrane. Through the open channels Na^+ flows into the cell and K^+ out of the cell giving rise to a net electrical current, an excitatory postsynaptic potential. This potential spreads through the entire cell membrane and causes muscle contraction (Hubbard, 1973). In the case of acetylcholine, an enzyme, acetylcholine esterase, destroys the neurotransmitter substance by splitting the ester back into its components, acetyl-CoA and choline. The enzyme acetylcholine esterase is made by the cholinergic neurons and also by muscle cells. In the nerve muscle synapse, the enzyme is not embedded in a muscle cell membrane but appears to be associated with a loose matrix of collagen and mucopolysaccharide fiber that crisscrosses the synaptic cleft and penetrates deep into the junctional folds. The speed with which the acetylcholine is destroyed is comparable to the rapid (millisecond) release. In fact, one-third is destroyed before it reaches receptor sites and the rest is hydrolyzed as it leaves the receptor. This enables the transmission process to be repeated several hundred times a second. Some snake venom toxicity can be related to the strong binding of such venoms (e.g., bungarotoxin) to the acetylcholine receptors (Chang and Lee, 1963). Drugs like physostigmine, on the other hand, inhibit acetylcholine esterase. Acetylcholine turnover is fairly rapid even in the absence of nerve impulses. Therefore, a supply of acetyl-CoA (glycolysis) and choline (plasma lecithin or choline) is needed at all times. The above-described process that occurs between the nerve terminal and the muscle cell is also applicable to signal transmission between the adjacent nerve cells.

9.3. DOPAMINE

Dopamine is a neurotransmitter that influences movement and behavior. Catecholamines when treated with formaldehyde vapor are converted to compounds that emit fluorescence. Therefore, fluorescence photomicrography, electron microscopy, and radioautography have aided

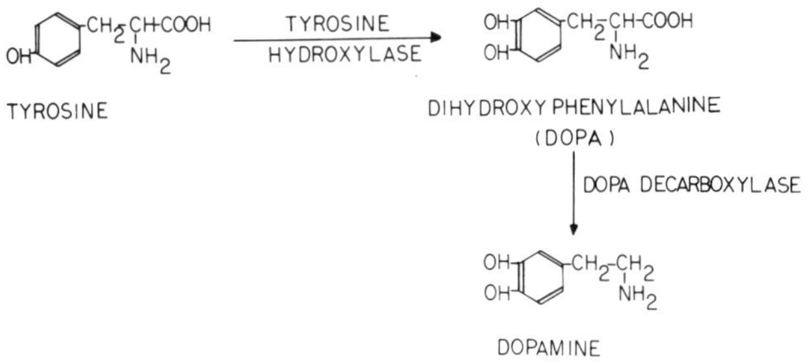

Figure 30.   Pathway of biosynthesis of dopamine. From Nagatsu *et al.* (1964).

in understanding the location and functions of dopaminergic neurons (Cooper *et al.*, 1974) (Fig. 30).

So far as the CNS is concerned, tyrosine is an essential amino acid. The manner in which dopamine is stored and released is very similar to that of acetylcholine discussed above. The destruction of dopamine, however, is quite different and handled without the aid of hydrolytic enzymes. Dopamine is metabolized by two enzymes, COMT and MAO (Axelrod, 1971) (Fig. 31). MAO is neuronal in origin and found in the mitochondria.

Regulation of levels of catecholamines is achieved by the following steps: when sympathetic nerves are stimulated, tyrosine hydroxylase activity increases but there is a feedback control (Axelrod, 1971). NA and dopamine themselves inhibit hydroxylase. Any increase in nerve-firing, due to stress, cold, or certain drugs, lowers the levels of catecholamines in the nerve terminals.

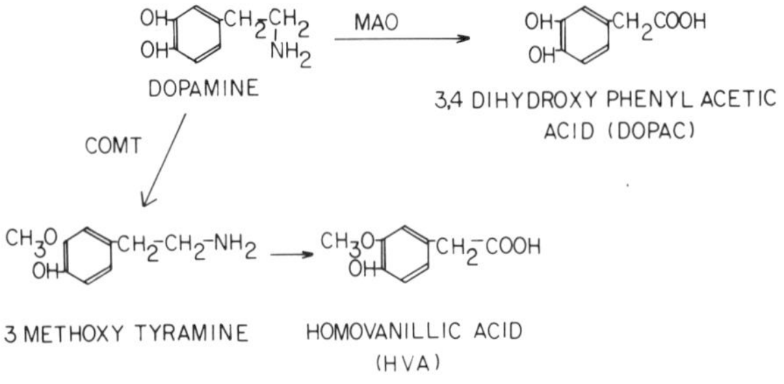

Figure 31.   Catabolism of dopamine.

9.3.1. Nutrition and Catecholamines

When rats were given large doses of tyrosine, there were no wide fluctuations in brain NA or dopamine levels. However, Gibson and Wurtman (1978) showed that tyrosine levels in the brain vary within a broad range in the fasting and postprandial periods. It was also noted that when brain levels of tyrosine levels were increased DOPA synthesis also increased in the brain (Wurtman *et al.*, 1974). In summary, one can say that when tyrosine levels are increased, the synthesis and release of dopamine is also elevated but this is quickly counteracted by a feedback mechanism that decreases tyrosine hydroxylase activity. As this is a rate-limiting step, metabolic equilibrium occurs.

The biochemical investigations on the synthesis and metabolism of dopamine led to a very important clinical method in alleviating symptoms of Parkinson's disease, a crippling disease in man. The original observation by Carlsson in Sweden in 1959 was an important clue. He found that when reserpine was given to rats it sharply reduced dopamine concentration in the dopaminergic neurons in the caudate nucleus in the brain and concurrently produced Parkinson-like tremors. Additionally, it was found that in patients of Parkinson's disease both in the caudate nucleus and putamen area the concentration of dopamine was significantly decreased (Sourkes *et al.*, 1976). This indicated that Parkinson's disease may be due to decreased levels of dopamine. Dopamine itself proved to be useless in therapy because of the impermeability to the blood–brain barrier. However, DOPA, the precursor of dopamine, was found to be readily taken up by the brain, following administration either intravenously or by the oral route. L-DOPA was far better than DL-DOPA and the amount needed for treatment was very much reduced. Although this dramatic result, mainly from the work of Cotzias and co-workers (1969), has helped a large number of patients of Parkinson's disease, it is obviously not a cure for the disease. After a long therapy, patients become refractory to the beneficial effects of L-DOPA probably due to the fact that fewer dopaminergic cells are available for converting L-DOPA to dopamine (Barbeau and Roy, 1976). Since the defect in Parkinson's disease is not in the decarboxylation step but rather in the hydroxylase step, L-DOPA is undoubtedly very useful in treatment. Since decarboxylation, however, can occur in other tissues, this could result in increased levels of dopamine in nonneuronal tissues. This ultimately means that a large number of therapeutic doses could be wasted or rendered useless. Thus, inhibitors of decarboxylase enzyme were tried, e.g., L-α-methyl-DOPA. These inhibitors decreased the decarboxylation reaction of DOPA in the peripheral tissues but did not affect the action in the CNS because the inhibitor was kept out of the brain by the blood–brain barrier. Therefore, a combination therapy is now a common practice (Yahr *et al.*, 1972). Initially, up to 90% of the patients respond to L-DOPA. However, the response is not permanent. Increasing disability in more than 50% of patients after 2–5 years and in about 90% after 10 years sets in. The increased disability is characterized by increased involuntary movements or dementia usually associated with diurnal oscillations, the "on–off" phenomenon (Liberman *et al.*, 1979). The on–off phenomenon refers to a sudden loss of effectiveness with an abrupt onset of akinesia ("off" effect) that may last minutes to hours, followed by an equally sudden return of effectiveness ("on" effect). This phenomenon can occur without an intercedent dose of L-DOPA.

9.4. SEROTONIN

Serotonin (5-hydroxytrytamine, 5HT) is a neurotransmitter which is intimately involved with neuropsychopharmacology. Its occurrence in serum was suspected due to its ability to cause rapid and powerful constriction of smooth muscles. Rapport *et al.* (1948) isolated 5HT from blood platelets. Soon after that, it was found to be present in the brain and further work showed that it was virtually confined to raphe nuclei, a distinct group of neurons (Amin *et al.*, 1954). The precursor for 5HT is dietary tryptophan, an essential amino acid, and is synthesized by the pathway shown in Fig. 32 (Page, 1968).

Tryptophan does not occur in large amounts in dietary proteins, and, therefore, supply of this amino acid is low. Although both tyrosine and tryptophan get hydroxylated on their route to form neurotransmit-

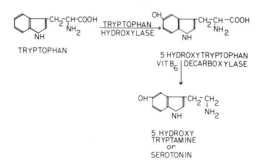

Figure 32.   Biosynthesis of 5HT.

ters, the enzymes are not similar. For example, tyrosine hydroxylase is a high-affinity enzyme and can operate efficiently even when substrate availability is low: In contrast, tryptophan hydroxylase is a low-affinity enzyme and, therefore, needs larger quantities of the substrate (Cohen and Wurtman, 1979). Since both dopamine and 5HT require amino acid precursors for synthesis, attention was drawn towards a relationship between dietary supply of the precursor amino acids and the neurotransmitter levels in the brain. Moir and Eccleston (1968) had established that precursor loading had a definite effect on the cerebral metabolism of 5-hydroxyindoles.

Fernstrom and Wurtman (1971a) reported that brain 5HT content depended upon the plasma tryptophan levels. If rats were fed tryptophan-poor diets (e.g., corn) the levels of brain 5HT were reduced (Fernstrom and Wurtman, 1971b). If, on the other hand, α-tryptophan was injected into rats receiving a tryptophan-poor diet there was a rapid repletion of brain serotonin (Fernstrom and Hirsch, 1975). Most surprisingly, the effect of the diet seemed to be very rapid; for example, in normal rats given a single meal of tryptophan-poor diet, within 2 hr there was a dramatic reduction in brain 5HT levels (Biggio et al., 1974). This led to the belief that neurotransmitter levels could be controlled by diet on a short-term basis. This new concept was contrary to earlier ideas that, at least in adulthood, the brain was capable of maintaining a rather constant level of nutrients (except glucose), irrespective of the nutrient level in the blood. Experiments proved that if tryptophan was injected into the bloodstream the normal levels of tryptophan and 5HT in the brain were elevated nine- and twofold, respectively.

Insulin is known to reduce blood amino acid levels but when this hormone was injected into the rats, surprisingly the blood level of tryptophan did not decrease, in fact it was elevated (Fernstrom and Wurtman, 1972). The same effect was observed after a high-carbohydrate diet (Fernstrom and Wurtman, 1971c). Undoubtedly the effect of the high-carbohydrate diet was mediated by the insulin response. Thus, it seemed that hormones are capable of regulating neurotransmitter levels.

To understand this tryptophan-sparing effect, one has to understand the underlying mechanism of amino acid transport into the brain. The transport of amino acids from the blood into the brain is carrier-mediated but the properties of the carriers for many of the different amino acids are not identical. Leucine, isoleucine, valine, phenylalanine, tyrosine, and tryptophan are transported by a common carrier protein. This means that all these amino acids will compete with each other for the same carrier protein (Blasberg and Lajtha, 1965). Only if the levels of the first five of the above amino acids are decreased can one expect a higher uptake of tryptophan. It was found that insulin or high-carbohydrate diet does decrease the levels of competing amino acids but it will not affect the tryptophan levels. One of the reasons for this difference may be due to the fact that tryptophan circulating in the blood is mostly bound to plasma proteins (albumin) with only about 10–20% circulating as unbound or free (McManamy and Oncley, 1958). Thus, the hormonal effect (insulin effect) did not change the bound tryptophan level but caused a decrease in the free tryptophan level.

As a further explanation of this hypothesis, Wurtman and co-workers point out that insulin has another physiological effect; it reduces lipolysis and consequently the FFA level in the plasma. Due to their decreased level, these FFA, normally bound to albumin, free some binding sites which are now available for binding tryptophan (Madras *et al.*, 1974). The ultimate effect of this process increases the bound tryptophan level in the blood and at the same time decreases the competing amino acids level. It is possible that the additional tryptophan is mobilized from tissues to bind with the vacant sites on the circulating plasma albumin.

The competition from other neutral amino acids for the uptake of tryptophan can also be seen in rats fed a high-protein diet. Such a diet will increase levels of all amino acids in the blood, including tryptophan, but this does not increase the tryptophan levels in the brain. In fact, this led to a decreased level in the brain. On the other hand, if a semisynthetic diet lacking in competitive amino acids is fed, the level of tryptophan (and 5HT) in the brain increased (Fernstrom *et al.*, 1975a). A similar experiment was tried with rats given a diet of skim milk (low fat), whole milk (moderate fat), and light cream (high fat) (Fernstrom *et al.*, 1975b). It was found that FFA levels changed in direct proportion to the fat content of the diet. When the dietary fat content was low (skim milk) serum FFA levels fell after food ingestion (insulin effect). When the diet was rich in fat, the higher influx of FFA negated the insulin effect. When FFA levels were high, bound tryptophan was low, but free tryptophan levels were high. In other words, the ratio of free to bound tryptophan increased. Under these conditions, it was found that there was no significant increase in the brain tryptophan level even though the free tryptophan level was high in the circulating plasma. Thus, the free tryptophan level was considered to be a poor indicator of the brain tryptophan level, under physiological conditions (as opposed to drug-induced conditions). This further confirmed that the ratio of bound tryptophan to total neutral amino acids in the circulating plasma determines the increase or decrease of brain tryptophan or 5HT levels rather than the blood level of tryptophan by itself (Fernstrom *et al.*, 1976).

9.4.1. Physiological Actions of 5HT

The most general effect of the iontophoretic application of 5HT is a reduction in excitability at least in the cortical, striatal, hypothalamic, and cerebellar neurons. In other areas, such as the thalamus, the action may be opposite (Phillis and Tebecis, 1967). 5HT may also play a role in sleep, sexual activity, and mood. Jouvet (1973) discovered that an electrolytic lesion of the raphe system caused insomnia in cats. It is already known that cell bodies of 5HT neurons are located in the raphe and reticular system of the brain stem. Studies in humans indicate that catecholamines inhibit and 5HT enhances REM sleep (Wyatt *et al.*, 1970). Administration of *p*-chlorophenylalanine, a potent inhibitor of 5HT action, produced hyperactivity (sexual) in rats (Shillito, 1970) but no such effect could be found in monkeys or humans.

9.5. CATABOLISM OF NEUROTRANSMITTERS: MONOAMINE OXIDASE

Monoamine oxidase (MAO) is an important enzyme in the catabolic breakdown of monoamine neurotransmitters (Costa and Sandler, 1972). Monoamine oxidase inhibitors were, therefore, sought for the treatment of disorders associated with neurotransmitter regulation (Figs. 33 and 34). Mental depression, for example, was treated with such drugs as iproniazine, a potent MAO inhibitor.

Iproniazine was tried in combination with isoniazid in the treatment of tuberculosis and was found to produce euphoria and help overcome depression. The mental depression, therefore was thought to be due to underproduction of monoamine neurotransmitters. Since MAO is a major enzyme that degrades such neurotransmitters, administration of an inhibitor of MAO would help to raise the level of monoamine neuro-

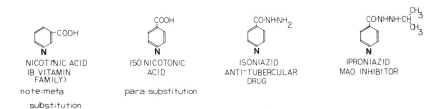

Figure 33. MAO inhibitors. Structural similarities with other known compounds.

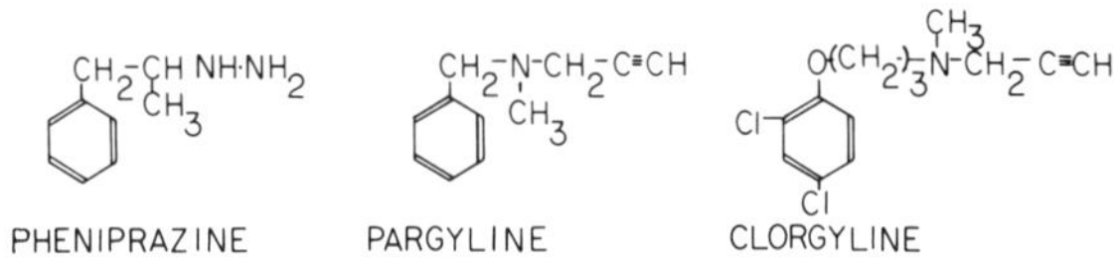

PHENIPRAZINE   PARGYLINE   CLORGYLINE

Figure 34.   MAO inhibitors.

transmitters which would ultimately help overcome depression (Loomer *et al.*, 1957).

Monoamine oxidase occurs in two forms, generally called a and b. The classical yellow enzyme present in outer membranes of mitochondria has flavine adenine dinucleotide FAD as a prosthetic group. An additional enzyme, called benzylamine oxidase, is pink and requires pyridoxal phosphate and Cu^{2+} ions. MAO is also useful in detoxification of amines formed by the intestinal flora. In general, the enzyme catalyzes the following reaction:

$$R-CH_2-CH_2-NH_2 + O_2 + H_2O \rightarrow RCH_2CHO + H_2O_2 + NH_3$$

The aldehyde can be further oxidized by an aldehyde dehydrogenase to form carboxylic acid.

Animals treated with MAO inhibitor show behavioral changes and are found to be alert and excited. Much can be learned from animal models but, as yet, therapy is generally viewed with caution. One side reaction needs to be considered in clinical use of MAO. When people have an abundance of cheese or milk products in their daily diets, they have an unusually high amount of tyramine in their systems (Asatoor *et al.*, 1963) (Fig. 35). Tyramine is normally oxidized by MAO to hydroxyphenylacetic acid and excreted in the urine (Fig. 36). When MAO inhibitor is given, tyramine accumulates. Tyramine is a "pressor" substance, i.e., it increases blood pressure. Subsequently, headache, vomiting, palpitation, and flushing are observed as side effects. The dietary item need not be restricted to milk products because even certain wines contain tyramine. In any case, administration of MAO inhibitors needs careful screening.

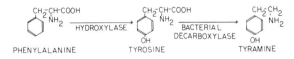

PHENYLALANINE   TYROSINE   TYRAMINE

Figure 35.   Formation of tyramine.

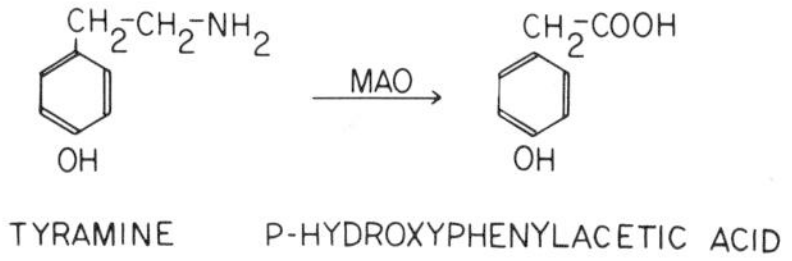

Figure 36. Normal catabolism of tyramine.

5HT is found in banana skins and may be responsible for the psychedelic effects when smoked (Page, 1968). Overuse of MAO inhibitors causes sedation and then, as 5HT builds up, a greatly excited state of mind. Later effects include ataxia or loss of muscular coordination and disorientation and unresponsive state of auditory or visual stimuli. These effects are similar to hallucinations experienced from ingestion of LSD. It is interesting to compare the chemical structures of some hallucinative drugs and 5HT (Page, 1968) (Fig. 37).

Numerous studies suggest that in chronic schizophrenia the blood platelet level of MAO is low. It has been shown that in eight patients of chronic paranoid schizophrenia the MAO activity in blood platelets was 4.81 nmole of benzaldehyde product per 10^8 platelets per hour, whereas in controls the activity was 12.3 nmole and in nonparanoid schizophrenics, 8.6 nmole (the benzaldehyde product was formed from a radioactive benzylamine used as a substrate). Later, when those patients who had an abnormal brain wave pattern, were mentally retarded, or were drug users or alcoholics were excluded from the group, the paranoid schizophrenic patients had a platelet MAO activity of 5.97 nmole (Potkin *et al.*, 1978).

Another hypothesis regarding biochemical alteration in schizophrenia assumes an increased level of NA in the brain (Ban, 1973). However, the administration of α-methyl-*p*-tyrosine, which is a specific inhibitor

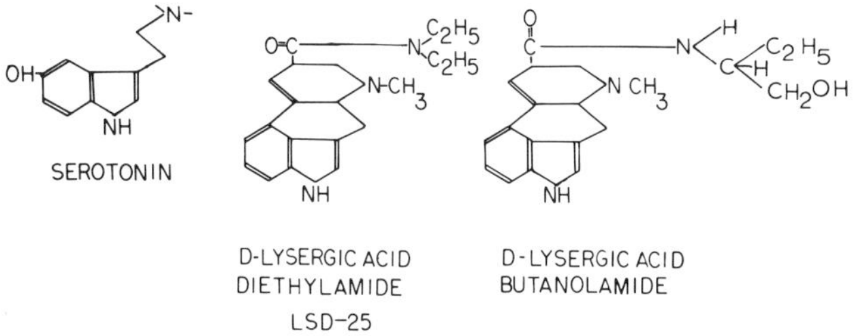

Figure 37. Similarities of structures of 5HT and some hallucinating drugs.

of tyrosine hydroxylase and reduces production of NA (without affecting dopamine synthesis), did not improve behavior in schizophrenic patients. Recently, it was reported that NA concentration in CSF is higher in patients with paranoid schizophrenia. It is possible that NA concentration in a specific area of the brain adjacent to cerebral ventricles is increased and spills over into the CSF. It is not known whether it is due to an abnormal increase in synthesis or a decrease in degradation (Lake *et al.*, 1980).

Yet another possibility was examined in the investigation of biochemical correlation between dopamine excess and schizophrenia. This hypothesis was supported by the fact that L-DOPA therapy produced aggravation of psychopathological symptoms in schizophrenic patients. Dopamine, however, may be related only to the behavioral aspects and not to the etiology of the complex disease. Finally, another hypothesis was proposed to explain some aspects of the disease. This hypothesis assumed that there was an abnormal "transmethylation" reaction in schizophrenics. For example, instead of physiological N-methylation of NA, O-methylation of dopamine occurs (Fig. 38). Feldstein (1970), however, found that the dimethoxy compound was not limited to schizophrenic individuals.

Another compound, adrenochrome, which is an oxidation product of epinephrine, was also suspected as the cause of psychopathological changes seen in schizophrenia. Altschule and Nayak (1971), however, failed to find an increased level of an oxidizing enzyme in the serum of schizophrenic patients.

Other theories regarding etiological factors of schizophrenia deal

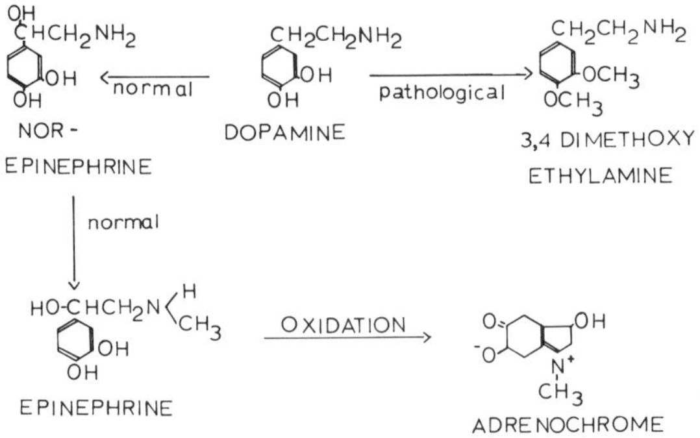

Figure 38. Normal and possible pathological pathways of dopamine metabolism.

with tryptophan–5HT metabolism (Kety, 1967). For example, there are two possible pathways of tryptophan metabolism:

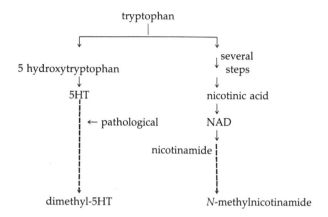

Interference with the formation of nicotinic acid will result in "excess" methyl groups because nicotinic acid is an efficient acceptor of methyl groups. The "excess" methyl group can then be diverted to form dimethyl-5HT. This compound is believed to be psychotoxic (Brune, 1967). But again the validity of this hypothesis was questioned when dimethyl-5HT was found in urine samples from normal healthy volunteers (Fourbye and Pind, 1964).

10

NEUROPEPTIDES

10.1. INTRODUCTION

Several years ago the discovery and understanding of the functions of neurotransmitters and their control of metabolic and behavioral aspects was one of the greatest achievements of neuroscientists. However, the recent discovery of neuropeptides has excited neuroscientists so much that it is hailed as the single most important breakthrough in our understanding of brain function. However, knowledge in this area is developing so rapidly that what is today an accepted concept may undergo a rapid change or modification in just a few years.

There are a number of peptides widely distributed throughout the nervous system, whereas there are others that are thought to be restricted to the hypothalamic–pituitary origin. Recently, even these were found to occur in other areas of the brain (Krieger, 1980). Those that are distributed throughout the CNS may be regarded as true neurotransmitters and the others as neurohormones that may act on the pituitary or peripheral organs and, therefore, are also called hypophysiotropic peptides. Examples of true brain neurotransmitter peptides distributed all through the CNS are (1) a decapeptide, substance P, (2) two pentapeptides, methionine enkephalin and leucine enkephalin, and (3) large peptides, endorphins. Neurotensin, somatostatin, and angiotensin II are other brain peptides of interest.

10.2. SUBSTANCE P

Substance P was discovered by von Euler and Gaddam in 1931 in extracts of horse intestine and brain that had an acetylcholinelike activity on smooth muscle cells but was not blocked by atropine (von Euler and Gaddam, 1931). Since it was later found to exist in higher amounts in the substantia nigra (Zetter, 1970), hypothalamus, basal ganglia, thalamus, and dorsal roots of the spinal cord and possessed acetylcholinelike properties, Lembek proposed a neurotransmitter role for this compound. Until 1970 not much progress was made to elucidate the chemical structure of this compound. In that year, Chang and Leeman (1970) accidentally found, in bovine hypothalamic extracts, a substance that stimulated secretion of saliva in rats. Again, as in the case of von Euler, they found this response was not blocked by atropine. The substance was identified as similar to substance P that von Euler had discovered and it consisted of 11 amino acids, Arg-Pro-Lys-Pro-Gln-Gln-Phe-Phe-Gly-Leu-Met (Leeman and Mroz, 1974).

Modern techniques of radioimmunoassay and immunohistochemical techniques (coupled with bovine gamma globulin) were used to find the concentration of substance P in the brain. The levels, ~50 pmol/g, were extremely low, compared to known neurotransmitters such as dopamine and serotonin. Substance P was found in nerve endings and released by a calcium-dependent process similar to other neurotransmitters (Schenker et al., 1976). To be a neurotransmitter it should be synthesized in the nerve endings and have a biochemical mechanism for rapid inactivation. The former condition was already proved but later studies have shown that the inactivation is readily achieved by an enzymatic (neutral endopeptidase) reaction. High-affinity uptake has not yet been demonstrated. The most important physiological action of this polypeptide is to initiate peristalsis of the gut. In any case, this is where it was originally found. The detailed role of substance P in the CNS is yet to be elucidated.

10.3. ENKEPHALINS

The search for new biologically active peptides continues even today. Goldstein and co-workers (1971) first reported that they found an association between levorphenol, a D(−) cogener of morphine, and brain tissue. This association of stereospecificity opened new vistas in the understanding of receptor sites and drug interactions. For instance, Pert and Snyder (1973), making use of a powerful inhibitor of morphine binding, naloxone, found that in subcellular fractions of brain homogenate the brain microsomal fraction had the greatest enrichment of opiate receptors, $\sim 32\%$, and the rest was distributed in mitochondrial and synaptosomal fractions. They also showed that the greatest amount of receptor binding occurred in the brain and that no specific binding was detected in innervated and plexus-free muscle. Similarly, no opiate receptor binding could be detected in human RBC. Regional distribution studies showed that striatum had the greatest enrichment of opiate receptors. A physiological concentration of calcium inhibits and chelating agents, such as ethylenediaminetetracetic acid and citrate, enhance receptor binding. This suggested that endogenous Ca^{2+} plays a role in receptor binding which is similar to that found in neurotransmitter action. Although this breakthrough goes a long way in our understanding of drug action and enables synthetic organic chemists and pharmacologists to prepare, test, and select both agonists and antagonists, it also helps immensely in treatment of hitherto untreatable mental diseases.

A small change in chemical structure can alter the physiological response immensely; for example, morphine is converted to heroin by acetylating the hydroxyl groups (Fig. 39). This change produced a more powerful drug, and heroin is known to enter the brain more easily. Some other compounds like naloxone can block the specific binding and, therefore, are called antagonists. Certain other compounds are both agonists and antagonists, e.g., nalorphine. Again the change in structure is relatively simple and minor (Fig. 40).

Some drugs containing equal portions of agonist and antagonist such as pentazocine are preferable because they are less addicting. The obvious use of an antagonist in medical practice can be appreciated when a small dose of naloxone or an antagonist can save a person from an overdose of morphine.

The greatest benefit from these studies came when researchers tried to answer the question, what are opiate receptors doing in the normal physiology of the brain? Why should nature provide in the brain such specific receptor proteins to drugs like morphine, undoubtedly foreign

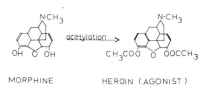

Figure 39. Simple chemical alteration increases biopotency dramatically.

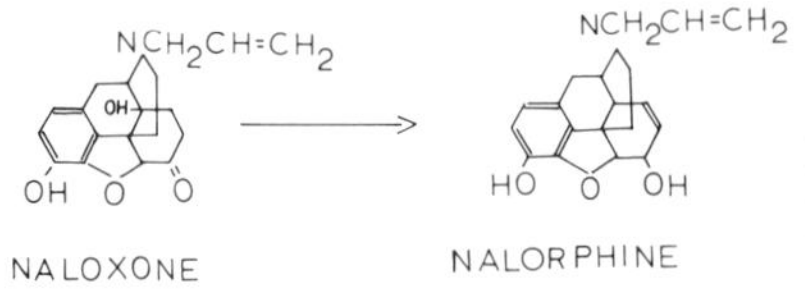

Figure 40. Simple chemical change produces profound difference in pharmacological actions.

to normal brain tissue. The obvious answer was that there may be substances in the brain that mimic the drug action. This idea was first proposed by Collier (1972). Hughes *et al.* (1975a), using brain extracts, suggested that the morphinelike substance was a low-molecular-weight peptide. Later, work by Hughes *et al.* (1975b) established the structures of two pentapeptides, methionine enkephalin and leucine enkephalin. This proved to be an astonishingly simple structure for such an important powerful agonist.

(N terminal) Tyr-Gly-Gly-Phe-Met   methionine enkephalin

(N terminal) Tyr-Gly-Gly-Phe-Leu   leucine enkephalin

Soon structural modifications were available for enhancing potency. For example,

Tyr-D-Ala-Gly-Me-Phe-Met(O)-ol

where MePhe = *N*-methylphenylalanine and Met(O)-ol = carboxyl of methionine changed to carbinol and the sulfur has been oxidized to sulfoxide, is a modified structure with increased potency (ability of peptide to inhibit binding of [³H]naloxine to receptors in rat brain) 28,800 times that of the original enkephalin. Use of D-alanine at the 2 position, replacing the glycine, prevented enzymatic degradation, and the most remarkable property of this compound was that it was effective even when it was given orally. This is the first case of an intact peptide ingested by mouth that reaches the brain. This new research produced a lot of excitement and kindled new hopes, but, most unfortunately, it was found that such polypeptides also produced physical dependence and increased tolerance, just like the analgesic drugs.

The physiological action of enkephalins is still not known in detail. However, iontophoretically applied Met-enkephalin depresses the firing rate of neurons in the cortex, neostriatum, and periaqueductal gray. This effect is similar to that of morphine and is inhibited by naloxine. Thus, the overall action seems to be one of inhibition or depression. The ideal peptide could be the one that has maximum analgesic but no addictive properties.

10.4. ENDORPHINS

All morphinelike compounds that occur endogenously in the brain and are polypeptide in nature but with a chain length greater than enkephalins were termed endorphins by Goldstein (1976). β-Lipotropin, a pituitary polypeptide discovered by Birk and Li (1964), was tested for opiatelike effects and was found to possess these properties. Further examination of this polypeptide revealed that β-lipotropin contained 91 amino acids, and units 61–65 were made of the same sequence of amino acids as Met-enkephalin. Three endorphins, α, β, and γ, were identified as fragments of β-lipotropin. The α fragment had the same sequence of 61 to 76, the β, 61 to 91, and the γ, 61 to 79 of the amino acids that exist in β-lipotropin. The β-endorphin is by far the most potent of the three. If this polypeptide is injected intravenously into rats, it produces prolonged muscular rigidity and immobility similar to schizophrenic catatonia (Bloom et al., 1976). Substantia gelatinosa in the spinal cord and lower brain stem is highly enriched in both enkephalin-containing neurons and opiate receptors (but not in β-endorphin neurons), and these may be involved in integrating the perception of pain. Naloxine can nullify hallucinations in schizophrenic patients, suggesting a role for brain peptides in schizophrenic delusions.

Although these findings raise hope, prolonged testing and research will have to be undertaken to exploit them for clinical applications. Other hormones isolated from anterior and posterior pituitary lobes like adrenocorticotrophic hormone, β-lipotropin, α-melanocyte-stimulating hormone, prolactin, growth hormone, and thyrotropin-stimulating hormone, prolactin, and thyrotropin-stimulating hormone have recently been found within the CNS (Krieger and Liotta, 1979) but not very much is known about their function.

In summary, brain peptides vary greatly in size; the smallest, a dipeptide, carnosine (L-histidine–β-alanine), seems to be localized primarily in olfactory neurons and the largest, β-lipotropin (91 amino acids), occurs mainly in the pituitary. Substance P, enkephalins, and other peptides have also been identified. All have potential in becoming key substances in the understanding of brain functions.

10.5. NUTRITIONAL SIGNIFICANCE

Dietary intake of foods containing proteins is known to provide the amino acids needed as building blocks for the synthesis of structural as well as functional (e.g., enzymes, hormones, neurotransmitters) proteins. This is one of the reasons why dietary protein is not only an obligatory requirement for mammals but the quality of protein is also a critical factor. For many years, it has been assumed that protein in the diet provides only a mixture of amino acids, because it was believed that protein is completely hydrolyzed to amino acids in the G.I. tract and that only free amino acids are absorbed into the circulation. If this were so, then preformed peptides, even if they were present in food or were formed by partial hydrolytic breakdown, would not be a source for brain peptides. However, there exists gluten enteropathy (celiac disease), the disease thought to be a result of ingestion of toxic peptides produced in the G.I. tract from dietary cereal protein such as wheat gluten (Frazer *et al.*, 1959). Similarly, there is a mental retardation associated with carnosinemia in which there is a deficiency of an enzyme that hydrolyzes the dipeptide, β-alanylhistidine (Scriver and Perry, 1972). These patients have elevated blood levels of carnosine (and associated anserine, a methylated form) and excrete increased quantities in urine. Thus, it seems possible that some peptides derived from dietary proteins can be absorbed without complete hydrolysis. One recalls similar observations in absorption of lipids. For many years it was thought that regardless of the form fed, labeled fatty acids appear in lymph, blood, and tissues almost entirely as triglycerides and phospholipids or sterol esters (Borgstrom, 1952a), and there was no evidence of absorption of unhydrolyzed ester of monohydric alcohols (Borgstrom, 1952b). However, evidence presented by Dhopeshwarkar and Mead (1962) clearly proved that orally fed methyl esters of fatty acids can be absorbed intact without hydrolysis.

Another possibility needs to be explored: in addition to intestinal absorption, there could be a rapid gastric absorption of polypeptides (Karel, 1948). If this can be established by current experiments and modern techniques, then the correlation between dietary proteins or partially hydrolyzed polypeptides and the origin of the brain polypeptides will be better understood. Even if all the above arguments are true, one has to explain why intestinal lymph or portal blood is almost free of polypeptides even after feeding high-protein diets (Sleisenger *et al.*, 1977). The possible explanation is that only those peptides that are resistant to hydrolysis by the intestinal enzymes can be absorbed without complete hydrolysis. Carnosine is an example of such a peptide and similar evidence has been reported by Warshaw *et al.* (1974). Even if one assumes that some polypeptides are absorbed intact into the bloodstream, whether they actually can get into the brain needs to be examined carefully due to the existence of the blood–brain barrier. Hemmings (1978) examined this question and concluded that biologically important peptides including α-gliadin (celiac disease) are not prevented from entering the brain when administered into the bloodstream. When the important brain peptide enkephalin was examined, Cornford et al. (1978) reported that only a small amount was taken up by the brain following intracarotid injection. In connection with the restriction posed by the blood–brain barrier there is some evidence to show that, at least with respect to some compounds, there is a greater uptake in infancy. It is also known that large amounts of immunoglobins are absorbed by the intestine without prior degradation by newborn babies. Thus, in infancy there may be a greater uptake of biologically active peptides. Such exorphins (peptides of exogenous origin) may include opiatelike peptides which induce drowsiness and sleep. Could this be a possible factor in drowsiness commonly observed in babies after feeding? Many interesting questions remain unanswered at this time, but more and more experiments tend to show a correlation between ingestion of food, partial hydrolysis of proteins, formation of exorphins, and finally uptake into the brain to produce biological effects. Not much is known about synthetic mechanisms to form these biologically active peptides in the brain itself. Considering the importance of the subject and number of laboratories that are actively seeking the answers, one can be sure that progress in this new and exciting field will be forthcoming very soon.

INBORN ERRORS
OF METABOLISM

11.1. LIPIDOSES

O'Brien has described the progress made in identifying lipid storage diseases as "enzymes to prevention." (O'Brien, 1973). The lipid-storage diseases, called "lipidoses," have been known since 1881 when Warren Tay, a British opthalmologist, together with Bernard Sachs, described a disease associated with the decline and finally arrest of cerebral development. Later on, this disease was named after the discoverers, Tay–Sachs disease. In later years, many diseases were reported in which there was an involvement of the CNS and also an abnormal accumulation of lipids. One exception to the above was found in Fabry's disease, where there was no involvement of the CNS, instead, the peripheral nervous system was affected. The accumulation of lipid may not always be in the brain; for example, in Gaucher's disease, the lipid accumulates both in the liver and the spleen. The nature of the biochemical lesion in diseases collectively called sphingolipidoses was traced to an accumulation of a complex lipid. The genesis and structure of these lipids is given below.

Step I

$$CH_3—(CH_2)_n—CH=CH—CH—CH—CH_2OH + CH_3(CH_2)_n—COOH \longrightarrow$$
$$\underset{OH\quad NH_2}{\big|\quad\big|}$$

Sphingosine   Fatty acid

$$CH_3—(CH_2)_n—CH=CH—CH—CH—CH_2OH$$
$$\underset{OH\quad NH—CO(CH_2)_n—CH_3}{\big|\quad\big|}$$

$\longrightarrow$

Acylsphingosine or ceramide (CER)

Step II

a. Ceramide + galactose$\longrightarrow$cerebroside

b. Ceramide + glucose$\longrightarrow$glucocerebroside

Step III: Formation of NANA

UDP-N-acetylglucosamine $\longrightarrow$ N- acetylmannosamine $\xoverset{ATP}{\longrightarrow}$

$\longrightarrow$ N-acetylmannosamine 6-phosphate

N-Acetylmannosamine 6-phosphate + phospho*enol*pyruvate $\xrightarrow{-P_i}$

$\longrightarrow$ N-acetylneuraminic acid 9-phosphate $\xrightarrow{-P_i}$ N-acetylneuraminic acid

N-Acetylneuraminic acid + CTP$\longrightarrow$CMP-N-acetylneuraminic acid + PP_i

Step IV: Formation of Gangliosides

CER—Glu—Gal—NANA	Ganglioside Gm_3
CER—Glu—Gal—NacGal	Ganglioside Gm_2
$\quad\quad\quad\quad\quad\big|$	
$\quad\quad\quad\quad$NANA	
CER—Glu—Gal—NAcGal—Gal	Ganglioside Gm_1
$\quad\quad\quad\quad\quad\quad\quad\quad\big|$	
$\quad\quad\quad\quad\quad\quad\quad$NANA	

A normal ganglioside contains four hexoses and is named Galb.

CER(1 $\leftarrow$ 1)Glu—(4 $\leftarrow$ 1)Gal—(4 $\leftarrow$ 1)NAcGal—(3 $\leftarrow$ 1)Gal
$\qquad\qquad\qquad\qquad\qquad\quad$3$\qquad\qquad\qquad\qquad\qquad\qquad\quad$3
$\qquad\qquad\qquad\qquad\qquad\quad\uparrow\qquad\qquad\qquad\qquad\qquad\qquad\quad\uparrow$
$\qquad\qquad\qquad$2 NANA—(8 $\leftarrow$ 2)NANA$\quad$ 2 NANA—(8 $\leftarrow$ 2)NANA

Several other complicated structures have been identified (numbers indicate bonding positions).

Starting from a base sphingosine (D-*erythro-trans*-2-amino-4-octadecene-1,3-diol, although an 18-carbon chain is the most common, it need not always be restricted to 18 carbons) and combining it with a fatty acid (to form an amide, ceramide), hexoses (to form cerebrosides and lactosides), and sialic acid, ganglioside GM_3 is formed. Finally, together with N-acetylgalactosamine and a number of sialic acid residues, the most complex sphingolipids are formed. Klenk and Svennerholm were responsible for identification and characterization of some of these complex lipids that accumulate in patients suffering from these forms of lipidoses.

The cause of the lipid accumulation could be either increased synthesis or decreased degradation. When experiments with tissue samples incubated with radioactive substrates were undertaken, it was found that the synthetic reaction was normal. For example, incubating spleen slices (obtained from autopsy material of patients with Gaucher's disease) together with $[1-^{14}C]$glucose or $[1-^{14}C]$ galactose produced no decrease or increase in the gluco- or galactocerebroside formed (Trams and Brady, 1960). Therefore, the possibility that a degradative pathway could be defective became obvious. Experiments using glucocerebroside confirmed that β-glucosidase, the enzyme that splits off glucose from glucocerebroside, was at a very low level in tissues of patients suffering from the infantile form of Gaucher's disease (Brady *et al.*, 1965). Similar studies in Niemann–Pick disease revealed a severe deficiency of sphingomyelinase (Brady *et al.*, 1966).

In the case of Tay–Sachs disease, the problem of deciding which degradative pathway is defective was not so simple because the accumulating lipid is much more complex. For example, it was evident that in Tay–Sachs disease ganglioside GM_2 accumulated in the brain. Two possibilities exist:

1. CER—Glu—Gal—GalNAc $\xrightarrow{\text{neuraminidase}}$ CER—Glu— Gal—GalNAc + NANA
        |
     NANA

2. CER—Glu—Gal—GalNAc $\xrightarrow{\text{hexoseaminidase}}$ CER—Glu—Gal + GalNAc
        | |
     NANA   NANA

Using an artificial fluorogenic substrate (4-methylumbelliferyl-β-D-N-acetylglucosaminide), it was found that the activity of hexosaminidase was actually increased severalfold rather than decreased! Similarly, the neuraminidase activity was also normal. This unexpected finding was soon clarified by Okada and O'Brien (1969). They found that there were two isozymes, hexosaminidases A and B, and only one of these, the A component, was deficient in tissues of patients suffering from Tay–Sachs disease. One question still remained to be answered. If neuraminidase is active in Tay–Sachs disease, one would have expected an asialo component (aminoglycolipid) to accumulate instead of GM_2. In fact, even this should not accumulate because hexosaminidase B, which degrades the asialo compound, is quite active in Tay–Sachs disease. Thus, there should be no accumulation of GM_2 at all. At present, it is believed that neuraminidase is not sufficiently active in neonatal life when ganglioside turnover is rapid. Another factor to be considered is that once Tay–Sachs ganglioside GM_2 starts to accumulate, it inhibits hexosaminidase B. Therefore, two simultaneous defects, hexosaminidase A deficiency and a very low or inhibited sialidase activity, lead to accumulation of GM_2. In any case, the absence of hexosaminidase A activity is considered as the primary defect in Tay–Sachs disease (Brady, 1972).

A list of well-documented lipid-storage diseases, lipids that accumulate, the enzyme defect, and some clinical features of the diseases is given in Table 6.

Having completed the investigation of the cause (enzyme defect) of these inherited diseases, the next question was to devise a simple test to diagnose the disease. The diagnosis had to be available to parents before birth of the child, otherwise (since there is no cure available yet) it would be too late. Some readily available patient material needs to be used for diagnostic tests. These would include leukocytes, skin fibroblasts, serum, urine, tears, or hair follicles. The substrates needed for enzyme assays should also be readily available and relatively inexpensive. It would be an added advantage to have the substrate release colored products for easy colorimetric determinations. Most of these criteria are now met with modern techniques and instruments (Saifer *et al.*, 1975). The largest program for identification of these genetic disorders undertaken in recent time has been to identify carriers of Tay–Sachs disease (Kaback *et al.*, 1977).

The lipid storage diseases are all transmitted as autosomal recessive disorders (except Fabry's disease, which is X-linked recessive) which means that when both parents are heterozygotes, statistically, there is one chance in four that pregnancy will result in a child born with the disease. Two children will be carriers and one may escape the disease altogether. In the case of Fabry's disease, only the female need be a

Table 6. Nature of Various Lipid-Storage Diseases

Disease	Accumulated lipid	Enzyme defect	Symptoms
Tay–Sachs	Ganglioside GM_2	Hexosaminidase A	Mental retardation; amaurosis; cherry red spot in macula
Fabry's	Ceramidetrihexoside	α-Galactosidase	Reddish purple maculopapular rash in umbilical, inguinal, and scrotal areas; renal impairment cornea opacities; no CNS involvement
Gaucher's	Glucocerebroside	β-Glucosidase	Mental retardation (infantile form only): hepatosplenomegaly; long bone involvement; lipid-laden cells in bone marrow
Niemann–Pick	Sphingomyelin	Sphingomyelinase	Generally similar to Gaucher's disease
Krabbe's	Increased ratio of galactocerebroside to sulfatide	β-Galactosidase	Mental retardation; globoid bodies in brain tissue
Metachromatic leukodystrophy	Sulfatide	Sulfatidase	Mental retardation; decreased nerve conduction time; metachromasia
Tay–Sachs variant	Globoside	Total hexosaminidase	Same as Tay-Sachs
Generalized gangliosidosis	GM_1	β-Galactosidase	Mental retardation; hepatomegaly; bone marrow involvement
Ceramide lactoside lipidoses	Lactosylceramide	β-Galactosidase	Slowly progressing CNS impairment; organomegaly; anemia

[a] Adapted from Brady (1978).

carrier. There is a 50 : 50 chance that her sons will be hemizygous and half of her daughters will be carriers. Thus, it is of utmost importance to know the prognosis of pregnancy as related to a chance of having a child born with the disease. For monitoring the risk, transabdominal amniocentesis is undertaken around the 14th gestational week and about 20 ml of the amniotic fluid is withdrawn. The fluid contains desquamated cells from the skin and mucous membrane of the fetus and a few of these cells survive in tissue culture growth medium. Their growth in about 3 to 4 weeks (more recently shorter periods) supplies sufficient numbers for enzyme assays. One must bear in mind the heavy burden the diagnostic procedure carries with it. It must be 100% correct or a life could be needlessly lost.

Although the prenatal diagnosis of lipid storage diseases has been immensely helpful to parents who then can be advised on termination of the pregnancy, so far, in spite of all the efforts, no definite cure has been found for these diseases. The cause of the disease is well established, but replacement of the missing enzymes or inducing enzyme activity has not yet been a complete success. The greatest problem is how to bypass the blood–brain barrier system and introduce a large protein (enzyme) into the brain where many of the lipids accumulate. Barranger *et al.* (1977) have reported a method which temporarily disrupts the blood–brain barrier (osmotic destruction) which then enables molecules as large as albumin (mol. wt. 68,000) to enter the brain. Male rats 300–350 g have been perfused with a hypermolar solution of D-mannitol into the right external carotid artery and the opening of the blood–brain barrier was monitored by penetration of Evans blue-albumin complex. Following enzyme infusion after opening the blood–brain barrier, the authors found an augmentation of α-mannosidase activity in the brain. There was no apparent gross neurological damage following this procedure. Assuming that an enzyme molecule (at least the size of α-mannosidase, ~180,000) crosses the brain capillaries, it still has to go from the extracellular space into the interior of the brain cell. Horseradish peroxidase has been shown to reach the glial and neuronal lysosomes, and so it is possible that the enzyme, once out of the capillaries, will ultimately reach the lysosomes. At present all this is still in the experimental stage and the techniques are not available in clinical treatment.

In the case of Gaucher's disease (adult form), older surviving patients seem to have escaped neurological damage and lipid accumulation in the CNS; instead the storage of lipids occurs in the spleen and the liver, so the injected material does not have to reach the brain. Some clinical trials have been initiated at National Institutes of Health. These workers report that following infusion of glucocerebrosidase (isolated and purified from human placental tissue), a 26% reduction in accumulated glucocerebroside was observed in two patients (Brady *et al.*, 1974). More impressive was the fact that the level of accumulated lipid not only came down but stayed normal over a period of months (Pentchew *et al.*, 1975). Thus, although many incidental problems need to be studied and worked out, there seems to be hope and enthusiasm for future research.

Organ grafts were also considered as substitute therapy, particularly during the early period when purified enzyme preparation was not available. A spleen was grafted into a patient with juvenile Gaucher's disease. However, not only was there no significant clinical improvement but complications of an immunological nature occurred (Groth *et al.*, 1971, 1972). Thus, this approach is also not without problems and complications. Only future research and study will reveal the success of this approach.

In the study of these diseases, one obviously needs an animal model for research and improvement. Chance mutation in animals may be one of the hopes of finding an animal suitable for this purpose, assuming that it reproduces similar genetic offspring. A cat with total hexosaminidase deficiency has been found (Cork *et al.*, 1977) and a canine with Krabbe's disease has also been reported (Fletcher *et al.*, 1977). Small animals, such as rats or mice, would be a definite advantage in terms of cost and time. Kanfer *et al.* (1975) demonstrated that an injection of conduritol-B-epoxide could inhibit β-glucosidase activity *in vivo*, and the level of glucosylceramide increased in several tissues. Adachi and Volk (1977) demonstrated that the intracellular inclusion bodies in the brain of injected mice were very similar to those found in the infantile form of the human disease. In another report Sukaragawa *et al.* (1977) mention that an injection of AY9944 [trans-1,4-bis(2-chlorobenzylaminomethyl) cyclohexane dihydrochloride] into neonatal rats causes a marked reduction of sphingomyelinase activity in the tissues of these animals. Further, there was almost a 300% increase in sphingomyelin of the liver in these rats following 17 days of treatment. This then can be an animal model for Niemann–Pick disease. Obviously, such chemically induced diseases cannot be transmitted to the next generation.

To summarize what we know about lipidoses:

1. The enzyme defects leading to accumulation of specific lipids, almost always accompanied by mental retardation and death in infancy or childhood, have been clearly identified.

2. Great advances have been made in diagnostic methods using readily available patient material, leukocytes, and skin fibroblasts.
3. Prenatal diagnosis has been a very successful and perfected procedure to advise parents about prognosis of disease in the offspring.
4. Although no cure has been found yet, research in animal models (developed via chemically induced enzyme deficiencies and chance mutation) and in clinical trials has been started in an attempt to replace the missing enzymes.

The future looks promising.

11.2. PHENYLKETONURIA

Phenylketonuria is one of the most prevalent (1 in 20–40,000 live births) of the aminoacidurias associated with mental retardation and, as such, has received much more attention than other similar genetic diseases. Nutrition plays an important role because if nutritional therapy is instituted early, most of the severe brain damage can be prevented. Folling (1934) came across a patient, a mother, whose two children were both mentally retarded. She also complained about a peculiar smell that always clung to her children. The usual urine analysis showed no abnormalities but a chance addition of ferric chloride to the urine produced a green color which was transient and disappeared a few minutes later. This gave Folling a clue. Several experiments later he isolated a substance that had a constant melting point of 155°C. The molecular weight was 164.4 and the substance was given an empirical formula $C_9H_8O_3$. Oxidation with $KMnO_4$ produced benzoic acid and oxalic acid. This suggested that the substance may be phenylpyruvic acid; authentic samples of this substance were identical to the compound isolated from the patients' urine and thus it was concluded that the children, who were mentally retarded were excreting phenylpyruvic acid (phenylketo acid) and the disease was named phenylketonuria.

The reason phenylpyruvic acid increased in the blood and also was excreted in the urine could be a defect in metabolism of the related essential amino acid, phenylalanine. Phenylalanine is normally hydroxylated to tyrosine in the body. Although such an enzyme, phenylalanine hydroxylase, does not occur in the brain, in the liver the reaction occurs readily. Other hydroxylases such as tyrosine hydroxylase and tryptophan hydroxylase do occur in the brain, but the hydroxylase reaction with phenylalanine is not so strong and may not have overall metabolic significance. The phenylalanine hydroxylase is an enzyme complex that needs an unconjugated pteridine cofactor and pyridine nucleotide-linked reductase (Craine *et al.*, 1972). The enzymatic activity seems to be stimulated by phospholipids (Fisher and Kaufman, 1972). Thus, either a total lack of this enzyme or reduced activity leads to accumulation of phenylalanine to as much as several hundred times the normal amount. This has serious consequences (Fig. 41). Some of the phenylalanine metabolites can be found in the brain. The apparent damage in the brain can be due to a breakdown of the transport system for the uptake of other aromatic essential amino acids (because of competitive inhibition by the high level of circulating phenylalanine in the blood). A decreased uptake of tyrosine or tryptophan can lead to deficiency of end products such as neurotransmitters. Other possible effects could be inhibition of obligatory enzyme systems such as pyruvic kinase and brain hexokinase (Weber, 1969). Phenylpyruvic acid is also shown to inhibit decarboxylation of DOPA (Boylen and Quastel, 1961). Failure in myelination is implied by Barbato *et al.*, (1968) and Shah *et al.* (1969) suggest interference in sterol synthesis.

The treatment of phenylketonuria has to begin very early, otherwise irreversible brain damage can occur. The treatment involves reducing phenylalanine in the diet (protein hydrolysate is passed through a column of charcoal and any amino acid that is removed is replaced by amino acids other than phenylalanine). Since caloric requirement and level of essential amino acids are very important for the growing infant, very careful monitoring of the blood levels and growth parameters are needed for successful treatment. It is important to realize that although nutritional therapy may not relieve all traces of the disease, the alternative is a definite mental retardation and associated problems.

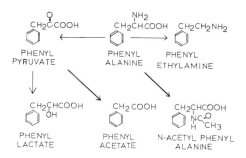

Figure 41. Metabolic transformations of phenylalanine.

11.3. MAPLE SYRUP URINE DISEASE

Maple syrup urine disease has been reported in countries all around the world, though it is not as common as the incidence of phenylketonuria. A survey by Snyderman (1967) in New York City showed its incidence as 1 in 60,000 births. The clinical features of this disease were first described by Menkes *et al.* (1954). Symptoms start within the first few days of life with poor appetite and a rapid progression of neurological symptoms. If the progression reaches semicomatose state then within the first weeks of life, death follows from respiratory arrest. A variant form of this disease is compatible with longer life. As early as the fifth day a characteristic odor, similar to the smell of maple syrup, is noticed in the urine and perspiration (hence the name of the disease). The biochemical aberration has now been elucidated and is related to branched-chain amino acid metabolism.

The reactions in Fig. 42 show the metabolic block. The first step in the metabolism of branched-chain amino acids, the transamination step, is not affected in this disease but the metabolic block occurs in the second step, i.e., decarboxylation. Due to this enzymatic failure, there is an accumulation of the respective keto acids, as well as their amino acid precursors. The diagnostic test consists of testing a urine sample with 2,4-dinitrophenyl hydrazine; a precipitate indicates the presence of keto acid (turbidity with pyruvic acid is a normal reaction). In another method, when leukocytes are incubated with [1-^{14}C] leucine, virtually no $^{14}CO_2$ is detected in specimens obtained from patients of this disease. Silberman *et al.* (1961) have reported a pathological examination of brain specimens and found a striking deficiency of myelin (lack of cerebrosides, sulfatides, and proteolipid protein) and, in some cases, astrocytosis.

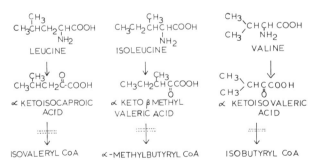

Figure 42. Metabolism of branched-chain amino acids. Note metabolic block in maple syrup urine disease.

Similar to the dietary treatment of phenylketonuria, a restriction of three amino acids in the diet, (much more difficult than a single amino acid) during very early stages, and strict monitoring of blood and urine levels is a recommended therapy.

It is always good to end with a note of optimism. In this book, an attempt has been made to touch on many aspects of brain development and to explain the biochemical basis of nutritional effects. However, a question that always haunts experts as well as laymen is whether all this knowledge will ultimately help in understanding the functioning of the brain both under physiological and pathological conditions. The answer to this question is based on very recent discoveries, some of which have already been cited in this monograph.

Until a few years ago, the visualization of an organ depended solely on X-ray techniques. With the use of radioisotope scanning, things that were invisible to the X-ray were observed on film or a screen. The introduction of computers in this field made the technique far more sophisticated and computed axial tomography gave cross-section views of organs in a living body. This was hailed as a milestone in diagnostic procedures. Now an even more powerful technique has emerged from work by Louis Sokoloff and a number of other researchers. This technique known as positron-emission tomography can look at the working of the cell in an intact organ in a living person. For example, following an introduction of a labeled compound, identical to the body's own metabolic product, the scanner visualizes areas that are actively utilizing this metabolite and separates them from areas where there may be a deficit in utilization. Thus, researchers at UCLA, led by Dr. David Kuhl, have discovered that during an epileptic seizure, a very active metabolic activity is going on in that area of the brain which has an electrical disturbance. This precise nature of the metabolic defect and the anatomical area associated with it is no doubt a key to ultimate understanding and a possible cure of many ailments that have eluded treatment for many years.

A basic understanding of the metabolic process and the effects of nutritional and environmental impacts on this process have taken a giant step in the direction of the ultimate goal of prevention and possible cure of debilitating and sometimes fatal diseases. The future looks very encouraging!

REFERENCES

Aaseth, J., 1973, The effect of mercaptodextran on distribution and toxicity of mercury in mice, *Acta Pharmacol. Toxicol.* **32**:430.

Abraham, E. P., 1939, Experiments relating to the constitution of alloxazine-adenine-dinucleotide, *Biochem. J.* **33**:543.

Ackefors, H., 1971, Mercury pollution in Sweden with special reference to conditions in the water habitat, *Proc. R. Soc. London Ser. B* **177**:365.

Adachi, M., and Volk, B., 1977, Gaucher's disease in mice induced by conduritol B epoxide: Morphological features, *Arch. Pathol.* **101**:255.

Aghajanian, G. K., and Bloom, F. E., 1967, The formation of synaptic junctions in developing rat brain: A quantitative electron microscopic study, *Brain Res.* **6**:716.

Agrawal, H. C., Davis, J. M., and Himwich, W. A., 1966, Postnatal changes in free amino acid pool of rat brain, *J. Neurochem.* **13**:607.

Alexander, F. R., 1974, The uptake of lead by children in differing environments, *Environ. Health Perspect.* **7**:155.

Alfin-Slater, R. B., and Aftergood, L., 1979, Nutritional role of hydrogenated fats (in rats), *in* "Geometrical and Positional Fatty Acid Isomers" (E. A. Emken and H. J. Dutton, eds.), pp. 53–74, Am. Oil Chem. Soc., Champaign, Ill.

Alling, C., and Svennerholm, L., 1969, Concentration and fatty acid composition of cholesterol esters of human brain, *J. Neurochem.* **16**:751.

Alter, H., Zvaifler, N., and Rath, C., 1971, Interrelationship of rheumatoid arthritis, folic acid and aspirin, *Blood* **38**:405.

Altman, J., 1966, Autoradiographic and histological studies of postnatal neurogenesis. II. A longitudinal investigation of the kinetics, migration and transformation of cells incorporating tritiated thymidine in infant rats, with special reference to postnatal neurogenesis of some brain regions, *J. Comp. Neurol.* **128**:431.

Altman, J., and Das, G. D., 1966, Autoradiographic and histological studies of postnatal neurogenesis. I. A longitudinal investigation of the kinetics, migration and transformation of cells incorporating tritiated thymidine in neonate rats, with special reference to postnatal neurogenesis in some brain regions, *J. Comp. Neurol.* **126**:337.

Altschule, M. D., and Nayak, U., 1971, Epinephrine cyclizing enzyme in schizophrenic serum, *Dis. Nerv. Syst.* **32**:51.

Amin, A. H., Crawford, T. B., and Gaddum, J. H., 1954, The distribution of substance P and 5 hydroxytryptamine in the central nervous system of the dog, *J. Physiol.* **126**:596.

Amin-Zaki, L., Elhassani, S., Majeed, M. A., Clarkson, T. W., Doherty, R. A., and Greenwood, M., 1974, Intrauterine methyl mercury poisoning in Iraq, *Pediatrics* **54**:587.

Anderson, R. E., and Sperling, L., 1971, Lipids of ocular tissues. VII. Positional distribution of the fatty acids in the phospholipids of bovine retina rod outer segments, *Arch. Biochem. Biophys.* **144**:673.

Anderson, T. A., and Fomon, S. J., 1974, Vitamins, *in* "Infant Nutrition" (S. J. Fomon, ed.), 2nd ed., Saunders, Philadelphia.

Anderson, J. M., Milner, R. D. G., and Strich, S. J., 1967, Effects of neonatal hypoglycemia on the nervous system: A pathological study. *J. Neurol. Neurosurg. Psychiat.* **30**:295.

Anonymous, 1981, Presence of 1, 25-dihydroxyvitamin D₃ receptor in rat pituitary, *Nutr. Rev.* **39**:140.

Anonymous, 1982, Inhibition of zinc absorption by inorganic iron, *Nutrition Rev.* **40**:76.

Ansell, G. B., and Spanner, S., 1967, The metabolism of labeled ethanolamine in the brain of the rat *in vivo*, *J. Neurochem.* **14**:873.

Antuono, P., Taiuti, R., Amaducci, L., and Pepeu, G., 1979, Preliminary trials of phosphorylcholine in Huntington's chorea and senile dementia, *in* "Nutrition and Brain," Vol. 5 (A. Barbeau, J. H. Growdon and R. J. Wurtman, eds.), pp. 331–333, Raven Press, New York.

Arakawa, T., 1970, Congenital defects in folate metabolism, *Am. J. Med.* **48**:594.

Arakawa, T., Fujii, M., and Hayashi, T., 1967, Dilatation of cerebral ventricles of rat offsprings induced by 6 mercapto purine administration to dams, *Tohoku J. Expt. Med.* **91**:143.

Arakawa, T., Mizuno, T., Sakai, K., Chida, K., Watanabe, A., Ohara, K., and Coursin, D. B., 1969, Electroencephalographic frequency patterns of rats treated with aminopterin in early infancy, *Tohoku J. Exp. Med.* **97**:385.

Asatoor, A. M., Levi, A. J., and Milne, M. D., 1963, Tranylcypromine and cheese, *Lancet* **ii**:733.

Axelrod, J., 1971, Noradrenaline: Fate and control of its biosynthesis, *Science* **173**:598.

Bachelard, H. S., 1974, "Brain Biochemistry," Chapman & Hall, London.

Bailey, K. V., 1965, Quantity and comparison breast milk in some New Guinea populations, *J. Trop. Pediatr.* **11**:35.

Baker, H., Frank, O., Zetterman, R. K., Rajan, K. S., Howe, W. T., and Levy, C. M., 1975, Inability of chronic alcoholics with liver diseases to use food as a source of folates, thiamine and vitamin B₆, *Am. J. Clin. Nutr.* **28**:1377.

Bakir, F., Damluji, S. F., Amin-Zaki, L., Murtadha, M., Khalidi, A., Al-Rawi, N.Y., Tikriti, S., Dhahir, H. I., Clarkson, T. W., Smith, J. C., and Doherty, R. A., 1973, Methyl mercury poisoning in Iraq, *in* "An Interuniversity Report," *Science* **181**:230.

Balazs, R., 1970, Carbohydrate metabolism, *in* "Handbook of Neurochemistry," Vol. 3 (A. Lajtha, ed.), pp. 1–36, Plenum, New York.

Balazs, R., 1973, Effects of disturbing metabolic balance in the early postnatal period, on brain development, *in* "Inborn Errors of Metabolism" (F. A. Holmes and C. J. Van den Berg, eds.), pp. 33–53, Academic, New York.

Ban, T. A., 1973, "Recent Advances in Biology of Schizophrenia," Charles C Thomas, Springfield, Ill.

Banga, I., Ochoa, S., and Peters, R. A., 1939, Pyruvate oxidation in brain. IV. The active form of vitamin B_1 and the role of C_4 dicarboxylic acids, *Biochem. J.* **33**:1109.

Barbato, L., Barbato, I. W. M., and Hamanaka, A., 1968, The *in vivo* effect of high levels of phenylalanine on lipids and RNA of the developing rabbit brain, *Brain Res.* **7**:399.

Barbeau, A., and Roy, M., 1976, Six year results of treatment with levodopa plus benzerazide in Parkinson's disease, *Neurology.* **26**:399.

Barboriak, J. J., and Meade, R. C., 1970, Effect of alcohol on gastric emptying in man, *Am. J. Clin. Nutr.* **23**:1151.

Barltrop, D., and Killala, N. J. P., 1967, Faecal excretion of lead by children, *Lancet* **ii**:1017.

Barnes, N. D., Hull, D., Balgobin, L., and Gompertz, D., 1970, Biotin responsive propionicacidemia, *Lancet* **ii**:244.

Barranger, J. A., Pentchev, P. G., Rapaport, S. I., and Bardy, R. O., 1977, Augmentation of a brain lysosomal enzyme activity following enzyme infusion, with concomitant alteration of the blood–brain barrier, *Trans. Am. Neurol. Assoc.* **102**:10.

Basnayake, V., and Sinclair, H. M., 1956, The effect of deficiency of essential fatty acids upon the skin *in* "Biochemical Problems of Lipids" (G. Popjak and E. Le Breton, eds.), pp. 476–484, Butterworth, London.

Bayer, S. M., and McMurray, W. C., 1967, The metabolism of amino acids in developing rat brain, *J. Neurochem.* **14**:695.

Bayoumi, R. A., and Smith, W. R. D., 1972, Some effects of dietary vitamin B_6 deficiency on GABA metabolism in developing rat brain, *J. Neurochem.* **19**:1883.

Bean, W. B., Hodges, R. E., and Daum, K., 1955, Pantothenic deficiency induced in human subjects, *J. Clin. Invest.* **34**:1073.

Beare-Rogers, J. L., 1979, Partially hydrogenated rapeseed and marine oils, *in* "Geometrical and Positional Fatty Acid Isomers" (E. A. Emken and H. J. Dutton, eds.), pp. 131–149, Am. Oil Chem. Soc., Champaign, Ill.

Beare-Rogers, J. L., Nera, E. A., and Craig, B. M., 1972, Accumulation of cardiac fatty acids in rats fed synthesized oils containing C_{22} fatty acids, *Lipids* **7**:46.

Beer, C. T., and Quastel, J. H., 1958, The effects of aliphatic aldehydes on the respiration of rat brain cortex slices and rat brain mitochondria, *Can. J. Biochem. Physiol.* **36**:531.

Benjamins, J. A., and McKhann, G. M., Development, regeneration and aging, *in* "Basic Neurochemistry" (G. J. Siegal, R. W. Albers, R. Katzman, and B. W. Agranoff, eds.), 2nd ed., Little, Brown, Boston.

Bernsohn, J., and Spitz, F. J., 1974, Linoleic and linolenic acid dependency of some brain membrane bound enzymes after lipid deprivation in rats, *Biochem. Biophys. Res. Commun.* **57**:293.

Bernsohn, J., and Stephanides, L. M., 1967, Are dietary defects a cause of multiple sclerosis? There is some evidence that lack of certain FA during maturation of brain results in demyelination, *Nature (London)* **215**:821.

Beutler, E., 1972, Drug induced anemia, *Fed. Proc.* **31**:141.

Bibergeil, H., Godel, E., and Amendt, P., 1975, Diabetes and pregnancy: Early and late prognosis of children of diabetic mothers, *in* "Early Diabetes in Early Life" (R. A. Camerini-Davalos, and H. S. Cole, eds.), pp. 427–434, Academic, New York.

Biggio, G., Fadda, F., Fanni, P., Talgliamonte, A., and Gessa, G. L., 1974, Rapid depletion of serum tryptophan, brain tryptophan, serotonin and 5-hydroxyindole acetic acid by a tryptophan-free diet, *Life Sci.* **14**:1321.

Birk, Y., and Li, C. H., 1964, β lipotropin, *J. Biol. Chem.* **239**:1048.

Blasberg, R., and Lajtha, A., 1965, Substrate specificity of steady state amino acid transport in mouse brain slices, *Arch. Biochem. Biophys.* **112**:361.

Blass, J. P., Avigan, J., and Uhlendorf, B. W., 1970, A defect in pyruvate decarboxylase in a child with intermittent movement disorder, *J. Clin. Invest.* **49**:423.

Bloch, K., Berg, B. N., and Rittenberg, D., 1943, Biological conversion of cholesterol to cholic acid, *J. Biol. Chem.* **149**:511.

Bloom, F., Segal, D., Ling, N., and Guilleman, R., 1976, Endorphins: Profound behavioral effects in rats suggest new etiological factors in mental illness, *Science* **194**:630.

Blum, K., Geller, I., and Wallace, J. E., 1971, Interaction effects of ethanol and pyrazole in laboratory rodents, *Br. J. Pharmacol.* **43**:67.

Bolanowska, W., 1968, Distribution and excretion of triethyl lead in rats, *Br. J. Ind. Med.* **25**:203.

Borgstrom, B., 1952a, Action of pancreatic lipase on tryglycerides *in vivo* and *in vitro, Acta Physiol. Scand.* **25**:328.

Borgstrom, B., 1952b, Intestinal absorption of ethyl oleate in the rat, *Acta Physiol. Scand.* **25**:322.

Boylen, J. B., and Quastel, J. H., 1961, Effect of L phenylalanine and sodium phenyl pyruvate on the formation of adrenaline from tyrosine in adrenal medula *in vitro, Biochem. J.* **80**:644.

Brady, R. O., 1972, Genetics of abnormal lipid metabolism, *in* "Current Topics in Biochemistry" (C. B. Anfinsen, R. F. Goldberger, and A. N. Schechter, eds.), pp. 1–48, Academic, New York.

Brady, R. O., 1978, Spingolipidosis, *Annu. Rev. Biochem.* **47**:689.

Brady, R. O., Kanfer, J. N., and Shapiro, D., 1965, Metabolism of glucocerebrosides. II. Evidence of enzymatic deficiency in Gaucher's disease, *Biochem. Biophys. Res. Commun.* **18**:22.

Brady, R. O., Kanfer, J. N., Mock, M. B., and Fredrickson, D. S., 1966, The metabolism of sphingomyelin. II. Evidence of an enzymatic deficiency in Niemann-Pick disease, *Proc. Nat. Acad. Sci. USA* **55**:366.

Brady, R. O., Pentchev, P. G., Gal, A. E., Hibbert, S. R., Dekaban, A. S., 1974, Replacement therapy for inherited enzyme deficiency, *N. Engl. J. Med.* **291**:989.

Brierley, J. B., Brown, A. W., and Mendrum, B. S., 1971, The nature and timecourse of the neurological alterations resulting from oligaemia and hypoglycemia in the brain of Macacca mulatta, *Brain Res.* **25**:483.

Brown, D. R., 1975, Neonatal lead exposure in the rat: Decreased learning as a function of age and blood lead concentrations. *Toxicol. Appl. Pharmacol.* **32**:628.

Brown, R. E., 1965, Decreased brain weight in malnutrition and its implications, *East Afr. Med. J.* **11**:584.

Brune, G. G., 1967, Tryptophan metabolism in psychoses in amines and schizophrenia, *in* "Amines and Schizophrenia" (H. E. Himwich, S. S. Ketty, and J. R. Smythies, eds.), Pergamon, Oxford.

Bunge, R. P., 1968, Glial cells and the central myelin sheath, *Physiol. Rev.* **48**:197.

Bunyan, J., Green, J., Diplock, A. T., and Robinson, D., 1967, Lysosomal enzymes and vitamin E deficiency. I. Muscular dystrophy, encephalomalacia and exudative diathesis in chick, *Br. J. Nutr.* **21**:127.

Burch, H. B., Lowry, O. H., Padilla, A. M., and Combs, A. M., 1956, Effect of riboflavin deficiency and realimentation of flavin enzymes at tissues, *J. Biol. Chem.* **223**:29.

Burr, G. O., and Burr, M. M., 1930, On the nature and role of fatty acids essential in nutrition, *J. Biol. Chem.* **86**:587.

Cahill, G. F., 1970, Starvation in man, *N. Engl. J. Med.* **282**:668.

Campbell, S., 1974, Physical methods of assessing size at birth, *in* "Size at Birth," Ciba Foundation Symposium 27, (E. Wolvestholme, ed.), p. 275, Elsevier/*Excerpta Medica*, Amsterdam.

Carroll, R. E., 1966, The relationship of cadmium in the air to cardiovascular disease death rates, *J. Am. Med. Assoc.* **198**:267.

Castle, W. B., 1929, Observations on the etiologic relationship of achylia gastrica to pernicious anemia. I. The effect of administration to patients with pernicious anemia of the contents of normal human stomach recovered after the injection of beef muscle, *Am. J. Med. Sci.* **178**:748.

Cearley, J. E., and Colman, R. L., 1974, Cadmium toxicity and bioconcentration in largemouth bass and bluegill, *Bull. Environ. Cont. Toxicol.* **11**:146.

Cebak, V., and Najdanvic, R., 1965, Effect of undernutrition in early life on physical and mental development, *Arch. Dis. Child* **40**:532.

Champakam, S., Srikantia, S. G., and Gopalan, C., 1968, Kwashiorkor and mental development, *Am. J. Clin. Nutr.* **21**:844.

Chan, W. Y., Cushing, W., Coffman, M. A., and Rennert, O. M., 1980, Genetic expression of Wilson's diseases in cell culture. A diagnostic marker, *Science* **208**:299.

Chang, C. C., and Lee, C. Y., 1963, Isolation of neurotoxins from the venom of *Bungarus multicinctus* and their modes of neuromuscular blocking action, *Arch. Intn. Pharmacodyn. Ther.* **144**:241.

Chang, M. M., and Leeman, S. E., 1970, Isolation of a sialogic peptide from bovine hypothalamus tissue and its characterization as substance P, *J. Biol. Chem.* **245**:4784.

Chase, P. H., Marlow, R. A., Dabiere, C. S., and Welch, N. N., 1973, Hypoglycemia and brain development, *Pediatrics* **52**:513.

Chida, N., Hirono, H., and Arakawa, T., 1972, Effects of dietary deficiency on fatty acid composition of myelin cerebroside in growing animals, *Tohoku J. Exp. Med.* **108**:219.

Chin, J. H., and Goldstein, D. B., 1977, Drug tolerance in biomembranes: A spin label study of the effect of ethanol, *Science* **196**:684.

Chin, J. H., Parsons, L. M., and Goldstein, D. B., 1979, Increased cholesterol content of erythrocyte and brain membranes in ethanol tolerant mice, *Biochim. Biophys. Acta* **513**:358.

Clarkson, T. W., Small, H., Mich, M., and Norseth, T., 1973, Excretion and absorption of methyl mercury after polythiol resin treatment, *Arch. Environ. Health* **26**:173.

Clarren, S. K., and Smith, D. W., 1978, The fetal alcohol syndrome, *N. Engl. J. Med.* **298**:1063.

Clarren, S. K., Alvord, E. C., Jr., Sumi, S. M., Streissguth, A. P., and Smith, D. W., 1978, Brain malformations related to prenatal exposure to ethanol, *J. Pediatr.* **92**:64.

Clausen, J., 1969, The effect of vitamin A deficiency on myelination in the CNS of the rat, *Eur. J. Biochem.* **7**:575.

Clausen, J., and Möller, J., 1967, Allergic encephalomyelitis induced by brain antigen after deficiency in polyunsaturated fatty acids during myelination: Is multiple sclerosis a nutritive disorder? *Acta Neurol. Scand.* **43**:375.

Cohen, E. L., and Wurtman, R. J., 1976, Brain acetylcholine: Control by dietary choline, *Science* **191**:561.

Cohen, E. L., and Wurtman, R. J., 1979, Nutrition and brain neurotransmitters, *in* "Nutrition: Pre- and Postnatal Development" (M. Winick, ed.), pp. 103–132, Plenum, New York.

Collier, H. O., 1972, Pharmacological mechanisms of drug dependence, *in* "Pharmacology and the Future Man" Vol. I (G. H. Acheson, ed.), pp. 65–76, Karger, Basel.

Collins, A. C., Cashaw, J. L., and Davies, V. E., 1973, Dopamine derived tetrahydroisoquinoline alkaloids, inhibitors of neuroamine metabolism, *Biochem. Pharmacol.* **22**:2337.

Cooper, J. R., Bloom, F. E., and Roth, R. H., 1974, "The Biochemical Basis of Neuropharmacology," 2nd ed., Oxford Univ. Press, London.

Cork, L. C., Munnel, J. F., Lorenz, M. D., Murphy, J. U., Baker, H. J., and Rattazzi, M.
C., 1977, GM_2 ganglioside lysosomal storage disease in cats with β-hexosaminidase
dependency, *Science* **196**:1014.

Cornford, E. M., Braun, L. D., Crane, P. D., and Oldendorf, W. H., 1978, Blood–brain
barrier restriction of peptides and low uptake of enkephalins, *Endocrinology* **103**:1297.

Costa, E., and Sandler, M., eds., 1972, The monoamine oxidase: New vistas, *in* "Advances
in Biochemical Psychopharmacology," Vol. 5, Raven, New York.

Cotzias, G. C., Van Woert, M. H., and Schiffer, L. M., 1967, Aromatic amino acids and
modification of Parkinsonism, *N. Engl. J. Med.* **276**:374.

Cotzias, G. C., Papavasilion, P. S., and Gellene, R., 1969, Modification of Parkinsonism-
chronic treatment with L-DOPA, *N. Engl. J. Med.* **280**:337.

Cox, D. H., and Harris, D. L., 1960, Effect of excess dietary zinc on iron and copper in
the rat, *J. Nutr.* **70**:514.

Cox, W. M., Jr., and Mueller, A. J., 1937, The composition of milk from stock rats and
an apparatus for milking small laboratory animals, *J. Nutr.* **13**:249.

Cragg, B. G., 1972, The development of cortical synapses during starvation in the rat,
Brain **95**:143.

Craine, J. E., Hall, E. S., and Kaufman, S., 1972, The isolation and characterization of
dihydropteridine reductase from sheep liver, *J. Biol. Chem.* **247**:6082.

Crapper, D. R., Krishman, S. S., and Dalton, A. J., 1973, Brain aluminium distribution
in Alzheimer's disease and experimental neurofibrillary degeneration, *Science*, **108**:511.

Cravioto, J., DeLicardie, E. R., and Birch, H. G., 1966, Nutrition growth and neurointe-
grative development: An experimental ecologic study, *Pediatrics* **38**:319.

Crawford, M. A., and Sinclair, A. J., 1972, The limitations of whole tissue analysis to
define linolenic acid deficiency, *J. Nutrition*, **102**:1315.

Crawford, M. A., Sinclair, A. J., Msuya, P. M., and Munhambo, A., 1973, Structural lipids
and their polyenoic constituents in human milk, *in* "Dietary Lipids and Postnatal
Development" (C. Galli, G. Jacini, and A. Pecile, eds.), pp. 41–56, Raven, New York.

Dahlquist, G., Persson, U., and Persson, B., 1972, Activity of D-β hydroxy butyrate dehydrogenase in fetal, infant and adult rat brain and influence of starvation, *Biol.
Neonate* **20**:40.

Davis, I. A., Miller, J. H., Lemmi, C. A., and Thompson, J. C., 1965, Mechanism and
inhibition of alcohol-stimulated gastric secretion, *Surg. Forum* **16**:305.

Davis, K. L., Mohs, R. C., Tinklenberg, J. R., Pfefferbaum, A., Hollister, L. E., and Kopell,
B. S., 1978, Improvement of long term memory processes in normal humans, *Science*
201:272.

Davis, S. D., Nelson, T., and Shepard, T. H., 1970, Teratogenecity of vitamin B_6 deficiency:
Omphalocele, skeletal and neural defects and splenic hypoplasia, *Science* **169**:1320.

Davison, A. N., 1972, Biosynthesis of myelin sheath, *in* "Lipids, Malnutrition and the
Developing Brain," Ciba Foundation Symposium (K. Elliott and J. Knight, eds.), pp.
73–90, Elsevier, Amsterdam.

Davison, A. N., and Peters, A., 1970, "Myelination," Charles C Thomas, Springfield, Ill.

Davison, A. N., Dobbing, J., Morgan, R. S., and Payling Wright, G., 1959, Metabolism
of myelin: The persistence of $[4-^{14}C]$ cholesterol in the mammalian central nervous
system, *Lancet* **i**:658.

Davson, H., 1967, "Physiology of the Cerebrospinal Fluid," Little, Brown, Boston.

Deguchi, T., Arata, J., Nishizuka, Y., and Hayaishi, O., 1968, Studies on the biosynthesis of nicotinamide adenine dinucleotide in the brain, *Biochim. Biophys. Acta* **158**:382.
de Guglielmone, A. E. R., Soto, A. M., and Duvilansky, B. H., 1974, Neonatal under-nutrition and RNA synthesis in developing rat brain, *J. Neurochem.* **22**:529.
DeLuca, H. F., and Schnoes, H. K., 1976, Metabolism and mechanism of action of vitamin D, *Ann. Rev. Biochem.* **45**:631.
Detering, N., Collins, R. M., Jr., Hawkins, R. L., Ozand, P. T., and Karahasan, A., 1980, Comparative effects of ethanol and malnutrition on the development of catecholamine neurons: Changes in neurotransmitter levels, *J. Neurochem.* **34**:1587.
Deuel, H. J., 1951, The effect of fat level of the diet on general nutrition. VIII. The essential fatty acid content of margarines, shortenings, butter and cottonseed oil as determined by a new biological assay method, *J. Nutr.* **45**:535.
Dhopeshwarkar, G. A., 1981, Naturally occurring food toxicants: Toxic lipids, in "Progress in Lipid Research," Vol. 19 (R. T. Holman, ed.), pp. 107–118, Pergamon, London.
Dhopeshwarkar, G. A., and Mead, J. F., 1961, Role of oleic acid in the metabolism of essential fatty acids, *J. Am. Oil Chem. Soc.* **38**:297.
Dhopeshwarkar, G. A., and Mead, J. F., 1962, Evidence for occurrence of methyl esters in body and blood lipids, *Proc. Soc. Exp. Biol. Med.* **109**:425.
Dhopeshwarkar, G. A., and Mead, J. F., 1973, Uptake and transport of fatty acids into the brain and the role of the blood–brain barrier system, in "Advances in Lipid Re-search," Vol. 11 (R. Paoletti and D. Kritchersky, eds.), pp. 109–142, Academic, New York.
Dhopeshwarkar, G. A., and Mead, J. F., 1975, Age and lipids of the CNS: Lipid metabolism in the developing brain, in "Aging," Vol. I, "Clinical, Morphologic and Neurochemical Aspects in the Aging CNS" (H. Brody, D. Harman, and J. M. Ordy, eds.), pp. 119–132, Raven, New York.
Dhopeshwarkar, G. A., and Subramanian, C., 1976, Intracranial conversion of linoleic acid to arachidonic acid: Evidence for lack of Δ^8 desaturase in the brain, *J. Neurochem.* **26**:1175.
Dhopeshwarkar, G. A., and Subramanian, C., 1977, Lipogenesis in the developing brain from intracranially administered [1-^{14}C] acetate and [U-^{14}C] glucose, *Lipids* **12**:762.
Dhopeshwarkar, G. A., Trivedi, J. C., Kulkarni, B. S., Satoskar, R. S., and Lewis, R. A., 1956, The effect of vegetarianism and antibiotics upon proteins and vitamin B_{12} in the blood, *Br. J. Nutr.* **10**:105.
Dhopeshwarkar, G. A., Maier, R., and Mead, J. F., 1969, Incorporation of [1-^{14}C] acetate into the fatty acids of the developing brain, *Biochim. Biophys. Acta,* **187**,16.
Dhopeshwarkar, G. A., Subramanian, C., McConnel, D. H., and Mead, J. F., 1972, Fatty acid transport into the brain, *Biochim. Biophys. Acta* **255**:572.
Dhopeshwarkar, G. A., Subramanian, C., and Mead, J. D., 1973, Metabolism of 1,2-[1-^{14}C] dipalmitoyl phosphatidylcholine, in the developing brain, *Lipids* **8**:753.
Dhopeshwarkar, G. A., Subramanian, C., Elepano-Gan, M., and Mead, J. F., 1981, Lack of vitamin E in the diet leads to essential fatty acid (EFA) deficiency, XII International Congress of Nutrition, San Diego, Calif., Abstract No. 782.
Dickerson, J. W. T., Dobbing, J., and McCance, R. A., 1967, The effect of undernutrition on the postnatal development of the brain and cord in pigs, *Proc. R. Soc. London Ser.* B **166**:396.
Dierks-Ventling, C., and Cone, A. L., 1971, Acetoacetyl CoA thiolase in brain, liver and kidney during maturation of the rat, *Science* **172**:380.
Dietrich, R., and Erwin, V. G., 1975, Involvement of biogenic amine metabolism in ethanol addiction, *Fed. Proc.* **34**:1962.

Dillon, M. J., 1974, Mental retardation, megaloblastic anemia, methyl malonyl aciduria and abnormal homocysteine metabolism due to an error in vitamin B_{12} metabolism, *Clin. Sci.* **47**:43.

DiPaolo, R. V., and Newbern, P. M., 1974, Copper deficiency and myelination in central nervous system of newborn rat: Histological and biochemical studies, *in* "Trace Substances in Environmental Health," Vol. 7 (D. D. Kemphill, ed.), Univ. of Missouri Press, Columbia.

Dobbing, J., 1961, The blood–brain barrier, *Physiol. Rev.* **41**:130.

Dobbing, J., 1964, The influence of early nutrition on the development and myelination of the brain, *Proc. R. Soc. London Ser. B* **159**:503.

Dobbing, J., 1971, Undernutrition and the developing brain, *in* "Handbook of Neurochemistry," Vol. 6 (A. Lai, ed.), pp. 255–266, Plenum, New York.

Dobbing, J., 1972, Vulnerable periods of brain development, *in* "Lipids, Malnutrition and the Developing Brain, Ciba Foundation Symposium" (K. Elliott and J. Knight, eds.), pp. 9–29, Elsevier, Amsterdam.

Dobbing, J., and Sands, J., 1973, Quantitative growth and development of human brain, *Arch. Dis. Child* **48**:757.

Dobbing, J., and Smart, J. L., 1973, Early undernutrition, brain development and behavior, *in* "Ethology and Development" (S. A., Barnett, ed., pp. 16–36, Spastics Int. Med. Publ. with Heinmann Med. Books, London.

Dodge, P. R., Prensky, A. L., and Feigin, R. D., 1975, "Nutrition and the Developing Nervous System," Mosby, St. Louis, Mo.

Dreyfus, P. M., and Victor, M., 1961, Effect of thiamine deficiency on the central nervous system, *Am. J. Clin. Nutr.* **9**:414.

Drillien, C. M., 1970, The small-for-date infant: Etiology and prognosis, *Pediatr. Clin. N. Am.* **17**:9.

Dutton, H. J., 1979, Hydrogenation of fats and its significance, *in* "Geometric and Positional Fatty Acid Isomers" (E. A. Emken and H. J. Dutton, eds.), pp. 1–16, Amer. Oil Chem. Soc., Champaign, Ill.

Eayrs, J. T., and Goodhead, B., 1959, Postnatal development of the cerebral cortex in the rat, *J. Anat.* **93**:385.

Ebady, M. S., Russel, R. L., and McCoy, E. E., 1968, The inverse relationship between the activity of pyridoxal kinase and the level of biogenic amines in rabbit brain, *J. Neurochem.* **15**:659.

Edmond, J., 1974, Ketone bodies as precursors of sterols and fatty acids in the developing brain, *J. Biol. Chem.* **249**:72.

Ehrlich, P., 1882, Ueber provozierte Fluorescenzerscheinungen am Auge, *Dtsch. Med. Wochenschr.* **8**:21, 35, 54.

Einarson, L., and Telford, R., 1960, Effect of vitamin E deficiency on the CNS in various laboratory animals, *Biol. Skr. Dan. Vidensk. Selsk.* **11**:1.

Ellenbogen, L., 1981, Drug—nutrient interactions, *in* "Nutrition, 'Eighty," A. R. Liss, ed., pp. 19–28, Lederle Labs., Wayne, N. J.

Elvehjem, C. A., Madden, R. J., Strong, S. M., and Woolley, D. W., 1938, The isolation and identification of anti black-tongue factor, *J. Biol. Chem.* **123**:137.

Enig, M. G., Munn, R. J., and Keeney, M., 1978, Dietary fat and cancer trends, a critique, *Fed Proc.,* **37**:2215.

Etienne, P., Gauthier, S., Dastoor, D., Collier, B., and Ratner, J., 1979, Alzheimer's disease: Clinical effect of lecithin treatment *in* "Nutrition and the Brain" (A. Barbeau, J. H. Growdon, and R. J. Wurtman, eds.), pp. 389–396, Raven, New York.

Eto, Y., and Suzuki, K., 1972a, Cholesterol esters in developing rat brain: Concentration and fatty acid composition, *J. Neurochem.* **19**:106.

Eto, Y., and Suzuki, K., 1972b, Cholesterol esters in developing rat brain: Enzymes of cholesterol ester metabolism, *J. Neurochem.* **19**:117.

Evans, H. M., and Bishop, K. S., 1922, Relations between fertility and nutrition, II. The ovulation rhythm in the rat on inadequate nutritional regimes, *J. Metabolic Res.*, **1**:319.

Eylar, E. H., 1972, The structure and immunogenic properties of myelin, *Ann. N. Y. Acad. Sci.* **195**:481.

Fang, S. C., 1974, Induction of C-Hg cleavage enzymes of rat liver by dietary selenite, *Res. Commun. Chem. Pathol. Pharmacol.* **9**:597.

Feldstein, A., 1970, Biochemical aspects of schizophrenia and antipsychotic drugs, *in* "Clinical Handbook of Psychopharmacology," (A. DiMascio and R. I. Shader, eds.) Science House, New York.

Fernstrom, J. D., and Hirsch, M. J., 1975, Rapid repletion of brain serotonin in malnourished corn-fed rats following l-tryptophan injection, *Life Sci.* **17**:455.

Fernstrom, J. D., and Wurtman, R. J., 1971a, Effect of chronic corn consumption on serotonin content of rat brain, *Nature New Biol.* **234**:62.

Fernstrom, J. D., and Wurtman, R. J., 1971b, Brain serotonin content: Physiological dependence on plasma tryptophan levels, *Science* **173**:149.

Fernstrom, J. D., and Wurtman, R. J., 1971c, Brain serotonin content: Increase following ingestion of carbohydrate diet, *Science* **174**:1023.

Fernstrom, J. D., and Wurtman, R. J., 1972, Elevation of plasma tryptophan by insulin in the rat, *Metabolism* **21**:337.

Fernstrom, J. D., Faller, D. V., and Shabshelowitz, H., 1975a, Acute reduction of brain serotonin and 5-HIAA following food consumption. Correlation with the ratio of serum tryptophan to sum of competing neutral amino acids, *J. Neural. Transm.* **36**:113.

Fernstrom, J. D., Hirsch, M. J., Madras, B. K., and Sudarsky, L., 1975b, Effect of skim milk, whole milk and light cream on serum tryptophan binding and brain tryptophan concentrations in rats, *J. Nutr.* **105**:1359.

Fernstrom, J. D., Hirsch, M. J., and Faller, D. V., 1976, Tryptophan concentration in rat brain. Failure to correlate with free serum tryptophan or its ratio to the sum of other neutral amino acids, *Biochem. J.* **160**:589.

Fisher, D. B., and Kaufman, S., 1972, The stimulation of rat liver phenylalanine hydroxylase by phospholipids, *J. Biol. Chem.* **247**:2250.

Fitzhardinge, P. M., and Stevens, E. M., 1972, The small-for-date infant. II. Neurological and intellectual sequelae, *Pediatrics* **50**:50.

Fleischer, S., and Rouser, G., 1965, Lipids of subcellular particles, *J. Am. Oil Chem. Soc.* **42**:588.

Fletcher, T. F., Suzuki, K., and Martin, F. B., 1977, Galactocerebrosidase activity in canine globoid leukodystrophy, *Neurology* **27**:758.

Folling, A., 1934, Uber Ausscheidung von Phenylbrenztraubensaure in dem Harn als Stoffwechselanomalie in Verbindung mit Imbezellität, *Z. Physiol. Chem.* **227**:169.

Fomon, S., 1974, "Infant Nutrition," 2nd ed., Saunders, Philadelphia.

Ford, D. H., 1973, Selected changes in developing postnatal rat brain, *in* "Development and Aging in the Nervous System" (M. Rockstein, ed.), Academic, New York.

Forsander, O., and Raiha, N., 1960, Metabolites produced in the liver during alcohol oxidation, *J. Biol. Chem.* **235**:34.

Fourbye, A., and Pind, K., 1964, Investigation on the occurrence of the dopamine metabolite 3,4 dimethoxyphenyl ethylamine in the urine of schizophrenics, *Acta Psychiat. Scand.* **40**:240.

Frankel, E. P., Kitchens, R. L., and Johnson, J. M., 1973, The effect of vitamin B_{12} deprivation on the enzymes of fatty acid synthesis, *J. Biol. Chem.* **248**:7540.

Frazer, A. C., Fletcher, R. F., Ross, C. A. C., Shaw, B., Sammons, H. G., and Schneider, R., 1959, Gluten induced enteropathy. The effect of partially digested gluten, *Lancet* **ii:**252.

Frazer, J. F. D., and Huggett, A. S. G., 1970, The partition of nutrients between mother and conceptuses in pregnant rat, *J. Physiol.* **207:**783.

Freeman, J. J., Choi, R. L., and Jenden, D. J., 1975, Plasma choline: Its turnover and exchange with brain choline, *J. Neurochem.* **24:**729.

Freinkel, N., and Metzger, B. E., 1979, Pregnancy as a tissue culture experience. The critical implication of maternal metabolism for fetal development, *in* "Pregnancy Metabolism, Diabetes and the Fetus," CIBA Foundation Symposium, No. 63 (K. Elliott and M. O'Connor, eds.), pp. 3–28, *Excerpta Medica,* Amsterdam.

Friberg, L., 1948, Proteinuria and emphysema among workers exposed to cadmium and nickel dust in a storage battery plant, *Proc. Int. Congr. Ind. Med.* **9:**641.

Fulco, A. J., and Mead, J. F., 1959, Metabolism of essential fatty acids. VIII. Origin of 5,8,11-eicosatrienoic acid in fat deficient rat, *J. Biol. Chem.* **234:**1411.

Futterman, S., Downer, J. L., and Hendrickson, A., 1971, Effect of essential fatty acid deficiency on fatty acid composition, morphology and electroretinographic responses of the retina, *Invest. Opthalmol.* **10:**151.

Gabbiani, G., Baic, D., and Deziel, C., 1967, Toxicity of cadmium for the CNS, *Exp. Neurol.* **18:**154.

Gaitonde, M. K., and Richter, D., 1966, Changes with age in the utilization of glucose carbon in liver and brain, *J. Neurochem.* **13:**1309.

Galli, C., Agradi, E., and Paoletti, R., 1974, n-6 Pentaene: n-3 hexaene fatty acid ratio as an index of linoleic acid deficiency, *Biochim. Biophys. Acta* **369:**142.

Galli, C., Spagnuolo, C., Agradi, E., and Paoletti, R., 1976, Comparative effects of olive oil and other edible fats on brain structural lipids during development, *in* "Lipids," Vol. I, "Biochemistry" (R. Paoletti, G. Porcellati, and G. Jacini, eds.), pp. 237–243, Raven, New York.

Gandy, G., Jacobson, W., and Sidman, R., 1973, Inhibition of transmethylation reaction in the CNS. An experimental model for subacute combined degeneration of the cord, *J. Physiol.* **233:**IP.

Ganong, W. F., 1971, "Review of Medical Physiology," 444 pp., Lange Med. Publ., Los Altos, Calif.

Garcia-Buñuel, L., McDougal, D. S., Burch, H. B., Jones, E. M., and Tonhill, E., 1962, Oxidized and reduced pyridine nucleotide levels and enzyme activities in brain and liver of niacin deficient rats, *J. Neurochem.* **9:**589.

Geel, S. E., and Dreyfus, P. M., 1975, Brain lipid composition of immature thiamine deficient and undernourished rats, *J. Neurochem.* **24:**353.

Gibson, C. J., and Wurtman, R. J., 1978, Physiological control of brain norepinephrine synthesis by brain tyrosine concentration, *Life Sci.* **22:**1399.

Gibson, G. E., Blass, J. P., and Jenden, D. J., 1978, Measurement of acetylcholine turnover with glucose used as precursor: Evidence for compartmentation of glucose metabolism in brain, *J. Neurochem.* **30:**71.

Gitlin, D., Kumate, J., and Morales, A. C., 1965, Metabolism and maternal fetal transfer of human growth homrone in the pregnant woman at term, *J. Clin. Endocrinol. Metab.* **25:**1599.

Goldberger, J., and Wheeler, G. A., 1927, Pellagra-preventive action of the cow pea and common wheat germ, U. S. Public Health Service *Public Health Reports* **42:**2383.

Goldmann, E. E., 1913, Vitalfarbung am Zentralnervensystem. *Abh. Preuss. Akad. Wiss. Phys.-Math. Kl.* No. 1, 1.

Goldstein, A., 1976, Opioid peptides (endorphins) in pituitary and brain, *Science* **193**:1081.

Goldstein, A., Lowrey, L. I., and Pal, B. K., 1971, Stereospecific and nonspecific interaction of the morphine congener, levorphanol, in subcellular fractions in mouse brain, *Proc. Nat. Acad. Sci. USA* **68**:1742.

Goodman, D. S., 1980, Vitamin A metabolism, *Fed. Proc.* **39**:2716.

Goodman, L. S., and Gilman, A., 1975, "Pharmacological Basis of Therapeutics," 5th ed., 920 pp., Macmillan, New York.

Gopalan, C., and Shrikantia, S. G., 1960, Leucine and pellagra, *Lancet* **i**:954.

Green, B. H., 1954, The formation from the Schwann cell surface of myelin in the peripheral nerves of chick embryo, *Exp. Cell Res.* **7**:558.

Groth, C. G., Hagenfeldt, L., Dreborg, S., Loftstrom, B., Ockerman, P. A., Sammuelsson, K., Svennerholm, L., Werner, B., and Westberg, G., 1971, Splenic transplantation in a case of Gaucher's disease, *Lancet* **i**:1260.

Groth, C. G., Bergstrom, K., Collste, L., Egberg, N., Hogman, C., Holm, G., and Moller, E., 1972, Immunologic and plasma protein studies in a splenic homograph recipient, *Clin. Exp. Immunol.* **10**:359.

Grover, W. D., and Scrutton, M. C., 1975, Copper infusion therapy in trichopoliodystrophy, *J. Pediatr.* **86**:216.

Growdon, J. H., Hirsch, M. J., Wurtman, R. J., and Weiner, W., 1977, Oral choline administration to patients with tardive dyskinesia, *N. Engl. J. Med.* **297**:524.

Growdon, J. H., Gelenberg, A. J., Doller, J., Hirsch, M. J., and Wurtman, R. J., 1978, Lecithin can suppress tardive dyskinesia, *N. Engl. J. Med.* **298**:1029.

Halas, E. S., and Sandstead, H. H., 1975, Some effects of prenatal zinc deficiency on behavior of adult rat, *Pediatr. Res.* **9**:94.

Halas, E. S., Hanton, M. J., and Sandstead, H. H., 1975, Intrauterine nutrition and aggression, *Nature (London)*, **257**:221.

Halstead, J. A., and Smith, J. C., Jr., 1969, Nutritional effects of clay ingestion (geophasia) in rats, *Fed. Proc.* **28**:691.

Halstead, J. A., Ronaghy, H. A., Abadi, P., Haghshenass, M., Amirhakemi, G. H., Barakat, R. M., and Reinhold, J. G., 1972, Zinc deficiency in man. The Shiraz experiment, *Am. J. Med.* **53**:277.

Hambidge, K. M., Hambidge, C., Jacobs, M., and Baum, J. D., 1972, Low levels of zinc in hair, anorexia, poor growth and hypogeusia in children, *Pediatr. Res.* **6**:868.

Hamlyn, L. H., 1962, An electron microscope study of pyramidal neurons in the Ammon's Horn of the rabbit, *J. Anat.* **97**:189.

Hanson, J. W., Streissguth, A. P., and Smith, D. W., 1978, Effect of moderate alcohol consumption during pregnancy on fetal growth and morphogenesis, *J. Pediatr.* **92**:457.

Hashim, S. A., and Asfour, R. H., 1968, Tocopherol in infants fed diet rich in polyunsaturated fatty acids, *Am. J. Clin. Nutr.* **21**:7.

Haubrich, D. R., and Chippendale, T. J., 1977, Regulation of acetylcholine synthesis in nervous tissue, *Life Sci.* **20**:1465.

Hawkins, R. A., and Biebuyck, J. F., 1979, Ketone bodies are selectively used by individual brain regions, *Science* **205**:325.

Hawkins, R. A., Williamson, D. H., and Krebs, H. A., 1971, Ketone body utilization by adult and suckling rat brain *in vivo*, *Biochem. J.* **122**:13.

Haworth, J. C., and Vidyasagar, D., 1971, Hypoglycemia in the newborn, *Clin. Obstet. Gynecol.* **14**:821.

Heckers, H., Korner, M., Tuschen, T. W. L., and Melcher, F. W., 1977, Occurrence of individual trans-isomeric fatty acids in human myocardium jejunum and aorta in relation to different degrees of atherosclerosis, *Atherosclerosis* **28**:389.

Heikkila, R., Cohen, R. G., and Dembiec, D., 1971, Tetrahydroisoquinoline alkaloids: Uptake by rat brain homogenates and inhibition of catecholamine uptake, *J. Pharmacol. Exp. Ther.* **179**:250.

Hemmings, W. A., 1978, The entry into the brain of large molecules derived from dietary protein, *Proc. R. Soc. London Ser. B* **200**:175.

Hers, H. G., 1963, Alfaglucosidase deficiency in generalized glycogen storate disease (Pompe), *Biochem. J.* **86**:11.

Higgins, E. S., 1962, The effect of ethanol on GABA content of rat brain, *Biochem. Pharmacol.* **11**:394.

Hilker, D. M., Chan, K. C., Chen, R., and Smith, R. L., 1971, Antithiamine effects of tea. I. Temperature and pH dependence, *Nutr. Rep. Int.* **4**(4):223.

Hill, D. E., Wyers, R. E., Holt, A. B., Scott, R. E., and Cheek, D. B., 1971, Fetal growth retardation produced by experimental placental insufficiency in Rhesus monkey: Chemical composition of brain liver, muscle and carcass, *Biol. Neonate* **19**:68.

Hill, E. G., Johnson, S. B., and Holman, R. T., 1978, Intensification of essential fatty acid deficiency in the rat by dietary *trans* fatty acids, *J. Nutr.* **109**: 1759.

Himwich, H. E., 1973, Early studies of the developing brain, *in* "Biochemistry of the Developing Brain," Vol. I (W. Himwich, ed.), pp. 1–53, Marcel Dekker, New York.

Himwich, W. A., Peterson, J. C., and Allen, M. D., 1957, Hematoencephalic exchange as a function of age, *Neurology* **7**:705.

Hirota, Z., 1898, Ueber die durch die Milch der an Kakke (Beriberi) leidenden Frauen verursachte Krankheit der Sauglinge, *Cent. Inn. Med.* **19**:385.

Hirsch, M. J., and Wurtman, R. J., 1978, Lecithin consumption increases acetyl choline concentrations in rat brain and adrenal gland, *Science* **202**:233.

Hirsch, M. J., Growdon, J. H., and Wurtman, R. J., 1978, Relations between dietary choline intake, serum choline levels and various metabolic indices, *Metabolism* **27**:953.

Hodgkin, D. C., Pickworth, J., Robertson, J. H., Prosen, R. J., and White, J. G., 1955, The crystal structure of the hexacarboxylic acid derived from B_{12} and the molecular structure of the vitamin, *Nature* **176**:325.

Holman, R. T., 1960, The ratio of trienoic/tetraenoic acids in tissue lipids as a measure of essential fatty acid requirement, *J. Nutr.* **70**:405.

Holman, R. T., 1973, Essential fatty acid deficiency in humans, *in* "Dietary Lipids and Postnatal Development" (C. Galli, G. Jacini, and A. Pecile, eds.), pp. 127–143, Raven, New York.

Holt, L. E., Jr., 1972, Folic acid deficiency, *in* "Pediatrics" (H. L. Barnett and A. H. Einhorn, eds.), 15th ed., Appleton–Century–Crofts, New York.

Hommes, O. R., Obbens, E. A. M. T., and Wijffels, C. C. B., 1973, The epileptogenic activity of sodium folate and the blood–brain barrier in the rat, *J. Neurol. Sci.* **19**:63.

Horecker, B. L., and Smyrniotis, P. Z., 1953, The coenzyme function of thiamine pyrophosphate in pentose metabolism, *J. Am. Chem. Soc.* **75**:1009.

Howell, J. M., and Thompson, J. N., 1967, Lesions associated with the development of ataxia in chicks, *Br. J. Nutri.* **21**:741.

Hoyumpa, A., Middleton, H., Wilson, F., and Schenker, S., 1974, Dual system of thiamine transport: Characteristics and effects of ethanol, *Gastroenterology* **66**:714.

Hubbard, J. I., 1973, Microphysiology of vertebrate neuromuscular transmission, *Physiol. Rev.* **53**:674.

Hughes, J., Smith, J. W., Kosterliz, H. W., Fothergill, L. A., Morgan, B. A., and Morris, H. R., 1975a, Identification of methionine enkephalin structure, *Nature (London)*, **258**:577.

Hughes, J., Smith, T. W., Morgan, B. A., and Fothergill, I., 1975b, Purification and properties of enkephalin—The possible endogenous ligand for the morphine receptor, *Life Sci.* **16**:1753.

Hunter, R., Barnes, J., Oakeley, H. F., and Mathews, D. M., 1970, Toxicity of folic acid given in pharmacological doses to healthy volunteers, *Lancet* **i**:61.

Hurley, L. S., 1969, Zinc deficiency in the developing rat, *Am. J. Clin. Med.* **22**:1332.

Hyyppä, M., 1971, Hypothalamic monoamines and pineal dopamine during the sexual differentiation of the rat brain, *Experientia* **27**:336.

Ikeda, M., Isuji, H., Nakamura, S., Ichiyama, A., Nishizuka, Y., and Hayaishi, O., 1965. Studies on the biosynthesis of nicotinamide adenine nucleotide. II. The role of picolinic carboxylase in the biosynthesis of nicotinamide adenine dinucleotide from tryptophan in mammals, *J. Biol. Chem.* **240**:1395.

Illingworth, D. R., and Portman, O. W., 1972, The uptake and metabolism of plasma lysophosphatidylcholine *in vivo* by the brain of squirrel monkeys, *Biochem. J.* **130**:557.

Isreal, Y., Salazar, I., and Rosenmann, E., 1968, Inhibitory effect of alcohol on intestinal amino acid transport *in vivo* or *in vitro*, *J. Nutr.* **96**:499.

Itoh, T., and Quastel, J. H., 1970, Acetoacetate metabolism in infant and adult rat brain *in vitro*, *Biochem. J.* **116**:641.

Itokawa, Y., Schultz, R. A., and Cooper, J. R., 1972, Thiamine in nerve membranes, *Biochim. Biophys. Acta* **266**:293.

Jagannatha, H. M., and Sastry, P. S., 1981, Cholesterol-esterifying enzymes in developing rat brain, *J. Neurochem.* **36**:1352.

Jäger, F. C., 1975, Linoleic acid intake and vitamin E requirement, *in* "The Role of Fats in Human Nutrition" (A. J. Vergroesen, ed.), pp. 381–432, Academic Press, New York.

Jelliffe, D. B., 1954, Infant feeding among the Yoruba of Needan, *West Afr. Med. J.* **2**:114.

Jelliffe, D. B., and Jelliffe, E. F. P., 1979, Early infant nutrition, *in* "Nutrition Pre- and Postnatal Development" (M. Winick, ed.), pp. 229–259, Plenum, New York.

Jepson, J. B., 1972, Hartnup disease, *in* "Metabolic Basis of Inherited Diseases" (J. B. Stanbury, J. B. Wyngaarden, and D. S. Fredrickson, eds.), pp. 1486–1503, Mc-Graw–Hill, New York.

Johnson, L., Schaffer, D., and Boggs, T. R., 1974, The premature infant, vitamin E deficiency, and retrolental fibroplasia, *Am. J. Clin. Nutr.* **27**:1158.

Jolliffe, M., Bowman, K. M., Rosenblum, L. A., and Fein, H. D., 1940, Nicotinic acid deficiency encephalopathy, *J. Am. Med. Assoc.* **114**:307.

Jones, E. L., and Smith, W. T., 1971, Hypoglycemic brain damage in neonatal rat, *Clin. Dev. Med.* **39/40**:231.

Jouvet, M., 1973, Serotonin and sleep in the cat, *in* "Serotonin and Behavior" (J. Barchas and E. Usdin eds.), pp. 385–400, Academic, New York.

Kaback, M. M., Rimoin, D. L., and O'Brien, J. S., 1977, "Tay-Sachs Disease Screening and Prevention," Liss, New York.

Kabara, J. J., 1973, A critical review of brain cholesterol metabolism, *in* "Progress in Brain Research" (D. H. Ford, ed.), pp. 363–382, Elsevier, Amsterdam.

Kalant, H., 1975, Direct effects of ethanol on the nervous system, *Fed. Proc.* **34**:1930.

Kandera, J., Levi, G., and Lajtha, A., 1968, Control of cerebral metabolite levels. II. Amino acid uptake and levels in various areas of the rat brain, *Arch. Biochem. Biophys.* **126**:249.

Kanfer, J. N., Legler, G., Sullivan, L., Raghavan, S., and Mumford, R. A., 1975, The Gaucher mouse, *Biochem. Biophys. Res. Commun.* **67**:85.

Karel, L., 1948, Gastric absorption, *Physiol. Rev.* **28**:433.

Kark, R. A. P., Menon, N. K., and Kishikawa, 1977, Mental retardation and mercury, *in* "Research to Practice in Mental Retardation, Biochemical Aspects," Vol. III (P. Mittler, ed.), pp. 271–280, Univ. Park Press, Baltimore, Md.

Karney, R. I., and Dhopeshwarkar, G. A., 1978, Metabolism of all *trans* 9, 12 [1-^{14}C] octadecadienoic acid in the developing brain, *Biochim. Biophys. Acta* **531**:9.

Karney, R. I., and Dhopeshwarkar, G. A., 1979, *Trans* fatty acids: Positional specificity in brain lecithin, *Lipids* **14**:257.

Keilin, D., and Mann, T., 1940, Carbonic anhydrase. Purifications and nature of the enzyme, *Biochem. J.* **34**:1163.

Kern, E. L., 1970, Vitamin A deficiency and glycolipid sulfation, *J. Lipid Res.* **11**:248.

Kerr, G. R., Waisman, H. A., Allen, J. R., Wallace, J., and Scheffler, G., 1973, Malnutrition studies in Macacca mullatta. II. The effect on organ size and skeletal growth, *Am. J. Clin. Nutr.* **26**:620.

Kety, S. S., 1967, Current biochemical approaches to schizophrenia, *N. Engl. J. Med.* **276**:325.

Kinsella, J. E., 1979, Effects of *trans trans* linoleate on the concentration of prostaglandins and their precursors in the rat, *Prostaglandins* **17**:543.

Kishimoto, Y., Williams, M., Mosser, H. W., Hignite, C., and Bierman, K., 1973, Branch-chain and odd numbered fatty acids and aldehydes in the nervous system of a patient with deranged vitamin B_{12} metabolism, *J. Lipid Res.* **14**:69.

Klee, C. B., and Sokoloff, L., 1967, Changes in D(−) β hydroxy butyrate dehydrogenase activity during brain maturation in the rat, *J. Biol. Chem.* **242**:3880.

Koeppe, R. E., and Han, C. H., 1962, Concerning pyruvate metabolism in rat brain, *J. Biol. Chem.* **237**:1026.

Koeppe, R. E., O'Neal, R. M., and Han, C. H., 1964, Pyruvate decarboxylation in thiamine deficient brain, *J. Neurochem.* **11**:695.

Kopeloff, L. W., Barrera, S. E., and Kopeloff, N., 1942, Recurrent convulsive seizures in animals produced by immunologic and chemical means, *Am. J. Psychiat.* **98**:81.

Kozak, L. P., and Wells, W. W., 1971, Studies on metabolic determinants of D-galactose induced neurotoxicity in the chick, *J. Neurochem.* **18**:2217.

Kramer, J. K. G., Hulan, H. W., Mahadevan, S., Sauer, F. D., and Corner, A. H., 1975, Brassica Campestris var. Span. II. Cardiopathogenicity of fractions isolated from rapeseed oil when fed to rats, *Lipids* **10**:511.

Krieger, D. T., 1980, Pituitary hormones in the brain: What is their function? *Fed. Proc.* **39**:2937.

Krieger, D. T., and Liotta, A. S., 1979, Pituitary hormones in brain: Where, how and why, *Science* **205**:366.

Krista, K., and Mancllovic, B., 1974, Transport of lead 203, and calcium 47 from mother to offspring, *Arch. Environ. Health* **29**:28.

Krnjevic, K., 1974, Chemical nature of synaptic transmission in vetebrates, *Physiol. Rev.* **54**:418.

Kurihara, T., Nussbaum, J. L., and Mandel, P., 1970, 2′,3′-Cyclic nucleotide 3′-phosphohydrolase in brains of mutant mice with deficient myelination, *J. Neurochem.* **17**:933.

Kurtz, D. J., Levy, H., and Kanfer, J. N., 1972, Cerebral lipids and amino acids in vitamin B_6 deficient suckling rat, *J. Nutrition* **102**:291.

Lahti, R. A., and Majchrowicz, E., 1967, The effect of acetaldehyde on serotonin metabolism, *Life Sci.* **6**:1399.

Lajtha, A., and Toth, J., 1973, Perinatal changes in free amino acid pool of the brain in mice, *Brain Res.* **55**:238.

Lake, C. R., Sternberg, D. E., Van Kammen, D. P., Ballenger, J. C., Ziegler, M. G., Post, R. M., Kopin, I. J., and Bunney, W. E., 1980, Schizophrenia: Elevated cerebrospinal fluid norepinephrine, *Science* **207**:331.

Lall, S. P., and Slinger, S. J., 1973, Nutritional evaluation of rapeseed oils and rapeseed soapstocks for laying hens, *Poultry Sci.* **52**:1729.

Langworthy, O. R., 1931, Lesions of the central nervous system characteristic of pellagra, *Brain* **54**:291.

Lauria, D. B., Joselow, M. M., and Browder, A. A., 1972, The human toxicity of certain trace elements, *Ann. Int. Med.* **76**:307.

Le Blancq, W. D., and Dakshinamurty, K., 1975, Nonparallel changes in brain monoamines in pyridoxine deficient rat, *Can. Fed. Biol. Soc.* **18**:32.

Ledeen, R. W., Skrivanek, J. A., Tirri, L. J., Margolis, R. K., and Margolis, R. U., 1976, Gangliosides of neurons: Localization and Origin, *in* "Ganglioside Function: Biochemical and Pharmacological Implications" (G. Porcellati, B. Ceccaralli, and G. Tettamanti, eds.), pp. 83–103, Plenum, New York.

Lee, H. M., Wright, L. D., and McCormick, D. B., 1973, Metabolism in the rat of biotin injected intraperitoneally as an avidin-biotin complex, *Proc. Soc. Exp. Biol. Med.* **142**:439.

Leeman, S. E., and Mroz, E. A., 1974, Substance P, *Life Sci.* **15**:2033.

Lehrer, W. P., Jr., Wiese, A. C., and Moore, P. R., 1952, Biotin deficiency in suckling pigs, *J. Nutr.* **47**:203.

Lepow, M. L., Bruckman, L., Robine, R. A., Markowitz, S., Gillette, M., and Kapish, J., 1974, Role of airborne lead in increased body burden of lead in Hartford children, *Environ. Health Perspect.* **7**:99.

Levine, E., and Scicli, G., 1969, Brain barrier phenomena, *Brain Res.* **13**:1.

Levitt, M., Nixon, P. F., Pincus, J. H., and Bertino, J. R., 1971, Transport characteristics of folates in cerebrospinal fluid. A study utilizing doubly labeled 5-methyl tetrahydrofolate and 5-formyl tetrahydrofolate, *J. Clin. Invest.* **50**:1301.

Levy, H. L., Mudd, S. H., Schulman, J. D., Dreyfus, P. M., and Ables, R. H., 1970, A derangement in B_{12} metabolism associated with homocysteinuria, cystathioninemia and methyl malonyl aciduria, *Am. J. Med.* **48**:390.

Liang, P. H., Hie, T. T., Jan, O. H., and Glok, L. T., 1967, Evaluation of mental development in relation to early malnutrition, *Am. J. Clin. Nutr.* **20**:1290.

Liberman, A., Neophytides, A., Kupersmith, M., Casson, I., and Durso, R., 1979, Treatment of Parkinson's disease with dopamine agonists: A review, *Am. J. Med. Sci.* **278**:65.

Lindblad, B. S., Ljungquist, A., Gebre-Mehdin, M., and Rahimtoola, A. R. J., 1977, The Composition and Yield of Human Milk in Developing Countries, Swedish Nutrition Foundation Symposium on Food and Immunology.

Lloyd-Still, J. D., Wolff, P. H., Horwitz, I., and Schwachman, H., 1972, Studies on intellectual development after severe malnutrition in infancy in cystic fibrosis and other intestinal lesions, *in* "Prognosis for the Undernourished Surviving Child: Cystic Fibrosis Management and Nutritional Aspects," Vol. 2, (A. Chavez, ed.), pp. 357–364, Proc. 9th International Congress on Nutrition, Mexico.

Lockwood, E. A., and Bailey, E., 1971, The course of ketosis and the activity of key enzymes of ketogenesis and ketone body utilization during development of the post-natal rat, *Biochem. J.* **124**:249.

Loewi, O., 1960, An autobiographic sketch, *Perspect. Biol. Med.* **4**:3.

Loo, Y. H., 1972, Levels of B_6 vitamins and pyridoxal phosphokinase in rat brain during maturation, *J. Neurochem.* **19**:1835.

Loomer, H. P., Saunders, J. C., and Kline, N. S., 1957, A clinical and pharmacodynamic evaluation of iproniazid as a psychic energizer, *Psychiat. Res. Rep.* No. 8, 129–141.

Lubchenco, L. O., and Bard, H., 1971, Incidence of hypoglycemia in newborn infants classified by birth and gestational age, *Pediatrics* **47**:831.

Lucis, O. J., Lucis, R., and Shaikh, Z. A., 1972, Cadmium and zinc in pregnancy and lactation, *Arch. Environ. Health* **25**:14.

Lundquist, F., 1962, Production and utilization of free acetate in man, *Nature (London)* **193**:579.

Madhok, T. C., and DeLuca, H. F., 1979, Characteristics of the rat liver microsomal enzyme system converting cholecalciferol into 25 hydroxycholecalciferol. I. Evidence for the participation of cytochrome P-450, *Biochem. J.* **184**:491.

Madras, B. K., Cohen, E. L., Munro, H. N., and Wurtman, R. J., 1974, Elevation of serum free tryptophan but not brain tryptophan by serum non esterified fatty acids, *Adv. Biochem. Psychopharmacol.* **11**:143.

Magos, L., Tuffery, A. A., and Clarkson, T. W., 1964, Volatilisation of mercury by bacteria, *Br. J. Ind. Med.* **21**:294.

Mann, G. V., Watson, P. L., McNally, A., and Goddard, J., 1952, Primate nutrition. II. Riboflavin deficiency in cebus monkey and its diagnosis, *J. Nutr.* **47**:225.

Mansour, M. M., Dyer, N. C., Hoffman, L. H., Schulert, A. R., and Brill, A. B., 1973, Maternal fetal transfer of organic and inorganic mercury via placenta and milk, *Environ. Res.* **6**:479.

Martin, D. C., Martin, J. C., Streissguth, A. P., and Lund, A., 1978, Sucking frequency and amplitude in newborns as a function of maternal drinking and smoking, *in* "Currents in Alcoholism," Vol. V, (M. Galanter, ed.), Grune and Stratton, New York, pp. 359–366.

Martins, M., and Dhopeshwarkar, G. A., 1982, Effect of high fat-high protein (low carbohydrate) diet on lipid metabolism in the rat, *Nutr. Rep. Int.* **25**:921.

Massieu, G. H., Ortega, B. G., Syrquin, A., and Tuena, M., 1962, Free amino acids in brain and liver of deoxypyridoxine-treated mice subjected to insulin shock, *J. Neurochem.* **9**:143.

McCandles, D. W., and Schenker, S., 1968, Encephalopathy of thiamine deficiency; Studies of intracerebral mechanism, *J. Clin. Invest.* **47**:2268.

McCay, P. B., Gibson, D. D., and Hornbrook, K. R., 1981, Colutathione-dependent inhibition of lipid peroxidation by a soluble, heat-labile factor not glutathione peroxidase, *Fed. Proc.* **40**:199.

McCollum, E. R., and Davis, M., 1913, Necessity of certain lipids in diet during growth, *J. Biol. Chem.* **15**:167.

McCoy, K. E., and Weswig, P. H., 1969, Some selenium responses in the rat not related to vitamin E, *J. Nutr.* **98**:383.

McCutcheon, J. S., Umemura, T., Bhatnagar, M. K., and Walker, B. L., 1976, Cardiopathogenicity of rapeseed oils and oil blends differing in erucic, linoleic and linoleic acid content, *Lipids* **11**:545.

McDonald, C. E., and Bolman, W. M., 1975, Delayed idiosyncratic psychosis with diphenyl-hydantoin, *J. Am. Med. Assoc.* **231**:1063.

McGeer, P. L., Eccles, J. C., and McGeer, E. G., 1978, "Molecular Neurobiology of the Mammalian Brain," Plenum, New York.

McIllwain, H., and Bachelard, H. S., 1971, "Biochemistry and the Central Nervous System," 4th ed., Churchill/Livingstone, Edinburgh.

McKenna, M. C., and Campagnoni, A. T., 1979, Effect of pre- and postnatal essential fatty acid deficiency on brain development and myelination, *J. Nutr.* **109**:1195.

McLaughlin, A. I. G., Kazantzis, G., King, E., Teane, D., Porter, R. S., and Owen, R., 1962, Pulmonary fibrosis and encephalopathy associated with the inhalation of aluminum dust, *Brit. J. Ind. Med.* **19**:253.

McMenamy, R. H., and Oncley, J. L., 1958, Specific binding of tryptophan to serum albumin, *J. Biol. Chem.* **233**:1436.

Mead, J. F., 1980, Membrane lipid peroxidation and its prevention, *J. Am. Oil. Chem. Soc.* **57**:393.

Mead, J. F., and Fulco, J. F., 1976, "The Unsaturated and Polyunsaturated Fatty Acids in Health and Disease," Charles C Thomas, Springfield, Ill.

Mead, J. F., Gan-Elepano, M., and Hirahara, F., 1980, Initiation of peroxidation by nitrogen dioxide in natural and model membrane systems, *in* "Nitrogen Oxides and Their Effects on Health" (S. D. Lee, ed.), pp. 191–197, Ann Arbor Science Publ., Ann Arbor, Mich.

Menkes, J. H., Hirst, P. L., and Craig, J. M., 1954, A new syndrome: Progressive familial cerebral dysfunction with an unusual urinary substance, *Pediatrics* **14**:462.

Menkes, J. H., Alter, M., Steigleder, G. K., Weakly, D. R., and Sung, J. H., 1962, A sex linked recessive disorder with retardation of growth, peculiar hair, focal cerebral and cerebellar degeneration, *Pediatrics* **29**:764.

Menon, N. K., and Dhopeshwarkar, G. A., 1981, Essential fatty acid deficiency and lipid metabolism of the developing brain, *in* "Progress in Lipid Research" (R. T. Holman, ed.), Vol. 20, Pergamon, Oxford, pp. 129–134.

Menon, N. K., and Kark, R. A. P., 1976, Inhibition of oxidation in chronic alkyl mercury poisoning, *Trans. Am. Soc. Neurochem.* **7**:151.

Menon, N. K., Moore, C., and Dhopeshwarkar, G. A., 1981a, Effect of essential fatty acid deficiency on maternal, placental and fetal tissues, *J. Nutr.* **111**:1602.

Menon, N. K., Subramanian, C., and Dhopeshwarkar, G. A., 1981b, Increased requirement of linoleate in the presence of dietary abundance of oleate, *Nutr. Rep. Int.* **23**:1063.

Merat, A., and Dickerson, J. W. T., 1974, The effect of the severity and timing of malnutrition on brain gangliosides in the rat, *Biol. Neonate* **25**:158.

Meyer, S., Maikel, R. P., and Brodie, B. B., 1959, Kinetics of penetration of drugs and other foreign compounds into cerebrospinal fluid and brain, *J. Pharmacol.* **127**:205.

Miale, I. L., and Sidman, R. L., 1961, An autoradiographic analysis of histogenesis in the mouse cerebellum, *Exp. Neurol.* **4**:277.

Michaelson, I. A., and Sauerhoff, M. W., 1974, An improved model of lead induced brain dysfunction in the suckling rat, *Toxicol. Appl. Pharmacol.* **28**:88.

Michaelson, J. C., 1948, Mode of development of vascular system of retina, with some observations on its significance for certain retinal diseases, *Trans. Opthalmol. Soc. U. K.* **68**:137.

Mickel, H. S., 1975, Multiple sclerosis: A new hypothesis, *Perspect. Biol. Med.* **18**:363.

Miettinen, J. K., 1973, Absorption and elimination of dietary Hg^{++} and methyl mercury in man, *in* "Mercury, Mercurials, and Mercaptans" (M. W. Miller and T. W. Clarkson, eds.), pp. 233–243, Charles C Thomas, Springfield, Ill.

Millen, J. W., and Hess, A., 1958, The blood–brain barrier an experimental study with vital dyes, *Brain* **81**:248.

Miller, S. A., 1970, Nutrition in the neonatal development of protein metabolism, *Fed. Proc.* **29**:1497.

Miller, S. L., Benjamins, J. A., and Morrell, P., 1977, Metabolism of glycerophospholipids of myelin and microsomes in rat brain, *J. Biol. Chem.* **252**:4025.

Minot, G. R., and Murphy, W. P., 1926, Treatment of pernicious anemia by a special diet, *J. Am. Med. Assoc.* **87**:470.

Mitchell, H. K., Snell, E. E., and Williams, R. J., 1941, Concentration of "folic acid," *J. Am. Chem. Soc.* **63**:2284.

Moir, A. T. B., and Eccleston, D., 1968, The effect of precursor loading on cerebral metabolism of 5-hydroxy indoles, *J. Neurochem.* **15**:1093.

Moore, C. E., and Dhopeshwarkar, G. A., 1980, Placental transparent of *trans* fatty acids in the rat, *Lipids* **15**:1023.

Moore, C. E., and Dhopeshwarkar, G. A., 1981, Positional specificity of trans fatty acids in fetal lecithin, *Lipids* **16**:479.

Moore, T., 1957, "Vitamin A," Elsevier, Amsterdam.

Mowat, J. H., Hutchings, B. L., Angler, R. B., Stokstad, E. L. R., Booth, J. H., Waller, C. W., Semb, J. and Subba Row, Y., 1948, Pteroid derivatives. I. Pteroyl-α-glutamyl acid and pteroyl-α,γ-glutamyl glutamic acid, *J. Am. Chem. Soc.* **70**:1096.

Nagatsu, T., Levitt, M., and Udenfriend, S., 1964, Tyrosine hydroxylase: The initial step in norepinephrine biosynthesis, *J. Biol. Chem.* **239**:2910.

Needleman, P., Raz, A., Minkes, M. S., Ferrendelli, J. A., and Sprecher, H., 1979, Triene prostaglandins: Prostacyclin and thromboxane biosynthesis and unique biological properties, *Proc. Nat. Acad. Sci. USA* **76**:944.

Niiyama, Y., Kishi, K., and Inone, G., 1970, Effect of ovarian steroids on maintenance of pregnancy in rats fed diets devoid of one essential amino acid, *J. Nutr.* **100**:1461.

Nikaido, T., Austin, J., Truel, L., and Rinehart, R., 1973, Studies in aging of the brain, II. Biochemical analyses of the nervous system in Alzheimer patients, *Arch. Neurol.* **27**:549.

Noble, E. P., and Tewari, S., 1973, Protein and RNA metabolism in brains of mice following chronic alcohol consumption, *Ann. N. Y. Acad. Sci.* **215**:333.

Noguchi, T., Cantor, A. H., and Scot, M. L., 1973, Mode of action of selenium and vitamin E in prevention of exudative diathesis in chicks, *J. Nutr.* **103**:1502.

Norton, W. T., and Podulso, S. E., 1973, Myelination in rat brain: Method of myelin isolation, *J. Neurochem.* **21**:749.

O'Brien, J. S., 1973, Tay-Sachs disease: From enzyme to prevention, *Fed. Proc.* **32**:191.

O'Dell, B. L., 1976a, Copper, *in* "Present Knowledge in Nutrition" (D. M. Hegsted, C. O. Chichester, W. J. Darby, K. W. McNutt, R. M. Stalvey, and E. H. Stotz, eds.), 4th ed., pp. 302–309, Nutrition Foundation, New York.

O'Dell, B. L., 1976b, Biochemistry and physiology of copper in vertebrates, *in* "Trace Elements in Human Health and Disease," Vol. I (A. S. Prasad, ed.), pp. 391–413, Academic Press, New York.

O'Dell, B. L., and Savage, J. E., 1960, Effect of phytic acid on zinc availability, *Proc. Soc. Exp. Biol. Med.* **103**:304.

Ogata, M., Mendelson, J. H., and Mello, N. K., 1968, Electrolyte and osmolality in alcoholics during experimentally induced intoxication, *Psychosom. Med.* **30**:463.

Okada, S., and O'Brien, J. S., 1969, Tay-Sachs disease: generalized absence of beta-D-N-acetylhexosaminidase component, *Science* **165**:698.

Oldendorf, W. H., 1971, Brain uptake of radiolabeled amino acids, amines, hexoses after intracranial injection, *Am. J. Physiol.* **221**:1629.

Ordonez, L. A., and Wurtman, R. J., 1974, Folic acid deficiency and methyl group metabolism in rat brain: Effects of L-DOPA, *Arch. Biochem. Biophys.* **160**:372.

Ortzonsek, N., 1967, The activity of heme ferro-lyase in rat liver and bone marrow in experimental lead poisoning, *Int. Arch. Gewerbepathol.* **24**:66.

Owen, O. E., Morgan, A. P., Kemp, H. G., Sullivan, J. M., Herrera, M. G., and Cahill, G. F. J., 1967, Brain metabolism during fasting, *J. Clin. Invest.* **46**:1589.

Owman, C., and Rosengren, E., 1967, Dopamine formation in brain capillaries—An enzymatic blood–brain barrier, *J. Neurochem.* **14**:547.

Page, I. H., 1968, "Serotonin," Year Book Med. Publ., Chicago.

Page, M. A., Krebs, H. A., and Williamson, D. H., 1971, Activities of enzymes of ketone body utilization in brain and other tissues of suckling rats, *Biochem. J.* **121**:49.

Pakkala, S. G., Fillerup, D. L., and Mead, J. F., 1966, The very long chain fatty acids of human brain sphingomyelin, *Lipids* **1**:449.

Paoletti, R., and Galli, C., 1972, Effects of essential fatty acid deficiency on the CNS in the growing rat, *in* "Lipids, Malnutrition and the Developing Brain" Ciba Foundation Symposium (K. Elliot and J. Knight, eds.), pp. 121–132, Associated Publishers, Amsterdam.

Parisi, A. F., and Vallee, B. L., 1969, Zinc metalloenzymes: Characteristics and significance in biology and medicine, *Am. J. Clin. Nutr.* **22**:1222.

Patel, A. J., and Balazs, R., 1975, Effect of X-ray irradiation of the biochemical maturation of rat cerebellum: Metabolism of {^{14}C} glucose and {^{14}C} acetate, *Radiat. Res.* **62**:456.

Pederson, J., 1975, Goals and end points in management of diabetic pregnancy, *in* "Early Diabetes in Early Life" (R. A. Camerini-Davalos and H. S. Cole, eds.), pp. 381–391, Academic, New York.

Pederson, J., Molsted-Pederson, L., and Anderson, B., 1974, Assessors of fetal perinatal mortality in diabetic pregnancy: Analysis of 1,332 pregnancies in the Copenhagen Series 1946–1972, *Diabetes* **23**:302.

Pentchev, P. H., Pradi, R. O., Gal, A. E., and Hibbert, S. R., 1975, Replacement therapy for inherited enzyme deficiency. Sustained clearance of accumulated glucocerebroside in Gaucher's disease following infusion of purified glucocerebrosidase, *J. Mol. Med.* **1**:73.

Perl, D. D., and Brody, A. R., 1980, Alzheimer's Disease: X-ray spectrometric evidence of aluminum accumulation in neurofibrillary tangle-bearing neurons, *Science* **208**:297.

Persson, B., and Tunnel, R., 1971, Influence of environmental temperature and acidosis on lipid mobilization in the human infant during first 2 hours after birth, *Acta Pediatr. Scand.* **60**:385.

Pert C., and Snyder, S. H., 1973, Opiate receptor: Demonstration in nervous tissue, *Science* **179**:1011.

Pfeifer, J. J., and Lewis, R. D., 1979, Effects of vitamin B_{12} deprivation on phospholipid fatty acid pattern in liver and brain of rats fed high and low levels of linoleate in low methionine diets, *J. Nutr.* **109**:2160.

Phear, E. A., and Greenberg, D. M., 1957, Methylation of deoxyuridine, *J. Am. Chem. Soc.* **79**:3737.

Phillis, J. W., and Tebecis, A. K., 1967, The responsiveness of thalamic neurons to iontophoretically applied monoamines, *J. Physiol.* **192**:715.

Pincus, J. H., 1972, Subacute necrotizing encephalomyelopathy (Leigh's disease). A consideration of clinical features and etiology, *Dev. Med. Child Neurol.* **14**:87.

Potkin, S. G., Cannon, E. H., Murphy, D. L., and Wyatt, R. J., 1978, Are paranoid schizophrenics biologically different than other schizophrenics, *N. Engl. J. Med.* **298**:61.

Prasad, A. S., Halstead, J. A., and Nadimi, M., 1961, Syndrome of iron deficiency anemia, hepatosplenomegaly, hypogonadism, dwarfism and geophasia, *Am. J. Med.* **31**:532.

Prasad, A. S., Miale, A., Farid, Z., Sandstead, H. H., and Schulert, A. R., 1963, Zinc metabolism in patients with the syndrome of iron deficiency anemia, hepatosplenomegaly, dwarfism and hypogonadism, *J. Lab. Clin. Med.* **61**:537.

Quastel, J. H., 1939, Respiration in the central nervous system, *Physiol. Rev.* **19**:135.

Rajalakshmi, R., Ali, S. Z., and Ramakrishnan, C., 1967, Effects of inanition during neonatal period on discrimination learning and brain biochemistry in the albino rat, *J. Neurochem.* **14**:29.

Rapport, M. M., Gree, A. A., and Page, I. H., 1948, Serum vasoconstrictor (serotonin). IV. Isolation and characterization, *J. Biol. Chem.* **176**:1243.

Rastogi, R. B., Merali, Z., and Singhal, R. L., 1977, Cadmium alters behavior, and bio-synthetic capacity for catecholamine and serotonin in neonatal rat brain, *J. Neurochem.* **28:**789.

Rawat, S. K., 1975, Effect of ethanol on brain metabolism, *Adv. Exp. Med. Biol.* **56:**165.

Reid, M. E., and Briggs, G. M., 1954, Nutritional studies with guinea pig. II. Pantothenic acid, *J. Nutr.* **52:**507.

Research News, 1973, *Science* **181:**253.

Reynolds, E. H., 1968, Mental effects of anticonvulsants and folic acid metabolism, *Brain* **91:**197.

Reynolds, E. H., Gallagher, B. B., Mattson, R. H., Bowers, M., and Johnson, R. A., 1972, Relationship between serum and cerebrospinal fluid folate, *Nature (London)* **240:**155.

Reynolds, W. A., and Pitkin, R. M., 1975, Transplacental passage of methylmercury and its uptake by primate fetal tissue, *Proc. Soc. Exp. Biol. Med.* **148:**523.

Robles, E. A., Mazey, E., Halsted, C. H., and Schuster, M. M., 1974, Effect of alcohol on motility of small intestine, *Johns Hopkins Med. J.* **135:**17.

Rockstein, M. (ed.), "Development and Aging in the Nervous System," Academic, New York.

Rocquelin, G., and Cluzan, R., 1968, Rapeseed oil with high and low levels of erucic acid and physiological effects on the rat. I. Effects on growth, feeding efficiency and state of different organs, *Ann. Biol. Anim. Biochem. Biophys.* **8:**395. [In French]

Roe, D. A., 1974, Minireview: Effects of drugs on nutrition, *Life Sci.* **15:**1219.

Roe, D. A., 1979, "Clinical Nutrition for the Health Scientist," 82 pp., CRC Press, Boca Raton, Fl.

Ronaghy, H., Spirey-Fox, M. R., Garn, S. M., Isreal, H., Harp, A., Moe, P. G., Petrosian, A., and Halstead, J. A., 1969, Controlled zinc supplementation for malnourished school boys. A pilot experiment, *Am. J. Clin. Nutr.* **22:**1279.

Ronagry, H. S., Reinhold, J. G., Mahloudji, M., Ghavami, P., Spirey-Fox, M. R., and Halsted, J. A., 1974, Zinc supplementation of malnourished school boys in Iran. Increased growth and other effects, *Am. J. Clin. Nutr.* **27:**112.

Rosenblum, W. I., and Johnson, M. G., 1968, Neuropathologic changes produced in suckling mice by adding lead to the maternal diet, *Arch. Pathol.* **85:**640.

Rosso, P., Hormazabel, J., and Winick, M., 1970, Changes in brain weight, cholesterol, phospholipid and DNA content in marasmic children, *Am. J. Clin. Nutr.* **23:**1275.

Rozear, R., DeGroof, R., and Somjen, G., 1971, Effects of microiontophoretic administration of divalent metal ions on neurons of the CNS of cats, *J. Pharmacol. Exp. Ther.* **176:**109.

Sacktor, B., Wilson, J. E., and Tickert, C. G., 1966, Regulation of glycolysis in brain *in situ* during convulsions, *J. Biol. Chem.* **241:**5071.

Saifer, A., Parkhurst, G. W., and Amoroso, J., 1975, Automated differentiation and mea-surement of hexosaminidase isoenzymes in biological fluids and its application to pre- and postnatal detection of Tay-Sachs disease, *Clin. Chem.* **21:**334.

Sandstead, H. H., Prasad, A. S., Schulert, A. R., Farid, Z., Miale, A., Bassilly, S., and Darby, W. J., 1967, Human zinc deficiency, endocrine manifestations and response to treatment, *Am. J. Clin. Nutr.* **20:**422.

Sandstead, H. H., Terhune, M., Brady, R. N., Gillespie, D., and Holloway, W. L., 1971, Zinc deficiency: Brain DNA, protein and lipids and liver ribosomes and RNA poly-merase, *Clin. Res.* **19:**83.

Sarma, J. M. K., and Rao, K. S., 1974, Biochemical composition of different regions in brains of small-for-date infants, *J. Neurochem.* **22:**671.

Schenker, C., Mroz, E. A., and Leeman, C. E., 1976, Release of substance P from isolated nerve endings, *Nature (London)* **264**:790.

Schlaepfer, W. W., 1969, Experimental lead neuropathy a disease of the supporting cells in peripheral nervous system, *J. Neuropathol. Exp. Neurol.* **28**:401.

Schrijver, R. D., and Privett, O. S., 1981, The effect of dietary *trans* fatty acids on rat liver microsomal 6 and 9-acyl desaturase activities, The American Oil Chemists Society Annual Meeting, New Orleans, Abstract 133.

Schroeder, H. A., 1965, Cadmium as a factor in hypertension, *J. Chron. Dis.* **18**:647.

Schuberth, J., Sollenberg, J., Sundwall, A., and Sorbo, B., 1965, Determination of acetyl CoA in brain, *J. Neurochem.* **12**:451.

Schwarz, K., and Foltz, C. M., 1957, Selenium as an integral part of factor 3 against dietary necrotic liver degeneration, *J. Am. Chem. Soc.* **79**:3292.

Scott, M. L., 1980, Advances in our understanding of vitamin E?, *Fed. Proc.* **39**:2736.

Scriver, C. R., and Perry, T. L., 1972, Disorders of β alanine and carnosine metabolism, *in* "Metabolic Basis of Inherited Diseases" (J. B. Stanbury, J. B. Wyngaarden, and D. S. Fredrickson, eds.), pp. 446–490, McGraw–Hill, New York.

Scriver, C. R., Mackenzie, S., Cluw, C. L., and Delvin, E., 1971, Thiamine responsive maple-syrup-urine disease, *Lancet* **i**:310.

Shah, S. N., Peterson, N. A., and McKean, C. M., 1969, Inhibition of sterol synthesis *in vitro* by metabolites of phenylalanine, *Biochim. Biophys. Acta* **187**:236.

Shillito, E. E., 1970, The effect of p-chlorophenylalanine on social interaction of male rats, *Br. J. Pharmacol.* **38**(2):305.

Siakotos, A. N., Koppang, N., Youmans, S., and Bucana, C., 1974, Blood levels of α tocopherol in a disorder of lipid peroxidation: Batten's disease, *Am. J. Clin. Nutr.* **27**:1152.

Signorelli, A., 1976, Influence of physostigmine upon consolidation of memory in mice, *J. Comp. Physiol.* **99**:658.

Silbergeld, E. K., and Goldberg, A. M., 1975, Pharmacological and neurochemical investigations of lead induced hyperactivity, *Neuropharmacology* **14**:431.

Silberman, J., Dancis, J., and Feigin, I., 1961, Neuropathological observations in maple syrup urine disease, *Arch. Neurol.* **5**:351.

Sims, K. L., and Pitts, F. N., Jr., 1970, Brain glutamate decarboxylase: Changes in the developing rat brain, *J. Neurochem.* **17**:1607.

Singhal, R. L., and Merali, Z., 1979, Biochemical toxicity of cadmium, *in* "Cadmium Toxicity" (J. H. Mennear, ed.), pp. 61–112, Marcel Dekker, New York.

Sinnhuber, R. O., Castell, J. D., and Lee, D. J., 1972, Essential fatty acid requirement of rainbow trout, *Salmo gairdneri*, *Fed. Proc.* **31**:1436.

Sleisenger, M. H., Pellig, D., Burston, D., and Mathews, D. M., 1977, Amino acid concentrations in portal-venous plasma during absorption from the small intestine of the guinea pig of an amino acid mixture simulating casein and a partial enzymatic hydrolysate of casein, *Clin. Sci. Mol. Med.* **52**:259.

Slinger, S. J., 1977, Improving the nutritional properties of rapeseed, *J. Am. Oil Chem. Soc.* **54**:94a.

Smith, M. E., 1968, The turnover of myelin in the adult rat, *Biochem. Biophys. Acta.*, **164**:285.

Smith, S., Watts, R., and Dils, R., 1968, Quantitative gas liquid chromatographic analysis of rodent milk tryglycerides, *J. Lipid Res.* **9**:52.

Smith, S. G., and Lasater, T. E., 1945, A progressive paralysis in dogs cured with synthetic biotin, *Am. J. Physiol.* **144**:175.

Snyderman, S. E., 1967, The therapy of maple syrup urine disease, *Amer. Dis. Child.* **113**(1):68.

Sobotka, T. J., Cook, M. P., and Brodie, R. E., 1974, Neonatal malnutrition: Neurochemical, hormonal and behavioral manifestations, *Brain Res.* **65**:443.

Sourkes, T. L., 1976, Parkinson's disease and other disorders of the basal ganglia, in "Basic neurochemistry," (G. J. Siegel, R. W. Albers, R. Katzman, and B. W. Agranoff, eds.), p. 673, Little, Brown, Boston.

Sprecher, H., 1975, Interconversions of polyunsaturated fatty acids, *in* "The Essential Fatty Acids," Miles Symposium, pp. 29–43, University of Manitoba, Miles Laboratories, Ontario.

Standal, B. R., Kao-Chen, S. M., Yang, G. Y., and Char, D. F., 1972, Early changes in pyridoxine status of patients receiving isoniazid therapy, *Am. J. Clin. Nutr.* **27**:479.

Stein, Z., Susser, M., Saenger, G., and Marolla, F., 1972, Nutrition and mental performance, *Science* **178**:708.

Stein, Z., Susser, M., Saenger, G., and Morolla, F., 1975, "Famine and Human Development, The Dutch Hunger Winter of 1944–1945," Oxford Univ. Press, New York.

Steinwall, O., 1969, Brain uptake of ^{75}Se-selenomethionine after damage to blood-brain barrier by mercurous ions, *Acta Neurol. Scand.* **45**:362.

Stempak, J. G., 1965, Etiology of antenatal hydrocephalus induced by folic acid deficiency in the albino rat, *Anat. Rec.* **151**:287.

Sternberg, S. S., and Phillips, F. S., 1958, 6 Aminonicotinamide and acute degenerative changes in the CNS, *Science* **127**:644.

Stewart, C. N., Bhagavan, H. N., Coursin, D. B., and Dakshinamurty, K., 1966, Effect of biotin deficiency on escape and avoidance learning in rats, *J. Nutr.* **88**:427.

Stone, W. J., Warnock, L. G., and Wagner, C., 1975, Vitamin B_6 deficiency in uremia, *Am. J. Clin. Nutr.* **28**:950.

Stone, W. L., Farnsworth, C. C., and Dratz, E. A., 1979, A reinvestigation of the fatty acid content of bovine, rat and frog retinal rod outerlayer segments, *Exp. Eye Res.* **28**:387.

Stowe, H. D., Wilson, M., and Goyer, R. A., 1972, Clinical and morphologic effects of oral cadmium toxicity in rabbits, *Arch. Pathol.* **94**:389.

Streissguth, A. P., Herman, C. S., and Smith, D. W., 1978, Intelligence, behavior and dysmorphogenesis in the fetal alcohol syndrome: A report on 20 patients, *J. Pediatr.* **92**:363.

Streissguth, A. P., Landesman-Dwyer, S., Martin, J. C., and Smith, D. W., 1980, Teratogenic effects of alcohol in humans and laboratory animals, *Science* **209**:353.

Sukaragawa, N., Sukaragawa, M., Kuwabara, T., Pentchev, P. G., Barranger, J. A., and Brady, R. O., 1977, Niemann-Pick disease experimental model: Sphingomyelinase reduction induced by AY 9944, *Science* **196**:317.

Sun, G. Y., Winniczek, H., Go, J., and Sheng, S. L., 1975, Essential fatty acid deficiency: Metabolism of 20:3W9 and 22:3W9 of major phosphoglycerides in subcellular fractions of developing and mature mouse brain, *Lipids* **7**:365.

Suzuki, T., and Yoshino, Y., 1969, Effect of D. penicillamine on urinary excretion of mercury in 2 cases of methyl mercury poisoning, *Jap. J. Ind. Health* **11**:487.

Svennerholm, L., 1976, Interaction of cholera toxin and ganglioside G_{M1}, *in* "Ganglioside Function, Biochemical and Pharmacological Implications" (G. Porcellati, B. Ceccarelli, and D. Tettamanti, eds.), pp. 191–229, Plenum, New York.

Svennerholm, L., Alling, C., Bruce, A., Karlsson, I., and Sapia, O., 1972, Effect on offspring of maternal malnutrition in the rat, *in* "Lipids, Malnutrition and Brain Development," Ciba Foundation Symposium (K. Elliott and S. Knight, eds.), pp. 141–157, Elsevier, Amsterdam.

Swaiman, K. F., Daleiden, J. M., and Wolfe, R. N., 1970, The effect of food deprivation on enzyme activity in developing brain, *J. Neurochem.* **17**:1387.

Sydenstricker, V. P., 1942, Preliminary observation in "egg white injury" in man and its cure with biotin concentrate, *Science* **95**:176.

Tanaka, C., and Cooper, J. R., 1968, The fluorescent microscopic localization of thiamine in nervous tissue, *J. Histochem. Cytochem.* **16**:362.

Tanphaichitr, V., 1976, Thiamin, *in* "Present Knowledge in Nutrition" (D. M. Hegsted, C. O. Chichester, W. J. Darby, K. W. McNutt, R. M. Stalvey, and E. M. Stotz, eds.), 4th ed., pp. 141–148, Nutrition Foundation, Washington, D.C.

Tappel, A. L., 1974, Selenium-glutathione peroxidase and vitamin E, *Am. J. Clin. Nutr.* **27**:960.

Taylor, R. T., and Weissback, H., 1969, E coli B. N^5 methyl tetra hydrofolate-homocysteine cobalamine methyl transferase: Activation of S-adenosyl-L-methionine and the mechanism of methyl group transfer. *Arch. Biochem. Biophys.* **129**:745.

Tewari, S., and Noble, E. P., 1975, Chronic ethanol ingestion by rodents effect on brain RNA, *in* "Alcohol and Abnormal Protein Biosynthesis" (M. A. Rothschild, M. Oratz, and S. S. Schreiber, eds), pp. 421–448, Pergamon, New York.

Tewari, S., and Noble, E. P., 1977, Ethanol induced changes in properties of brain ribosomes, *in* "Alcohol and Aldehyde Metabolizing Systems III" (R. G. Thurman, J. R. Williamson, H. R. Droth, and B. Chance, eds.), pp. 613–624, Academic, New York.

Tews, J. K., 1969, Pyridoxine deficiency and brain amino acids, *in* "Vitamin B₆ in Metabolism of the Nervous System" (M. A. Kinsall, ed.), *Ann. N.Y. Acad. Sci.* **166**:74.

Theuer, R. C., 1972, Effect of oral contraceptive agents on vitamin and mineral needs: A review, *J. Reprod. Med.* **8**:13.

Thomas, H. M., and Blackfan, K. D., 1914, Recurrent meningitis due to lead in a child of 5 years, *Am. J. Dis. Child* **8**:377.

Thomasson, H. J., 1953, Biological standardization of essential fatty acids (a new method), *Int. Z. Vitaminforsch.* **25**:26.

Thomasson, H. J., 1961, Acidi Grassi Essenziale, *Riv. Ital. Sostanze Grasse* **12**:541.

Thompson, J. N., and Scott, M. L., 1969, Role of selenium in the nutrition of the chick, *J. Nutr.* **97**:355.

Thompson, J. N., Beare-Rogers, J. L., Erody, P., and Smith, D. C., 1973, Appraisal of human vitamin E requirement based on examination of individual meals and a composite Canadian diet, *Am. J. Clin. Nutr.* **26**:1349.

Tilden, J. T., and Cornblath, M., 1972, Succinyl CoA: 3-Keto-acid CoA transferase deficiency. A cause for ketoacidosis in infancy, *J. Clin. Invest.* **51**:493.

Tinoco, J., Williams, M. A., Hincenberg, I., and Lyman, R. L., 1971, Evidence for non-essentiality of linolenic acid in the diet of the rat, *J. Nutr.* **101**:937.

Tinoco, J., Miljanich, P., and Medwadowsky, B., 1977, Depletion of docosahexaenoic acid in retinal lipids of rats fed a linolenic acid-deficient, linoleic acid containing diet, *Biochim. Biophys. Acta* **486**:575.

Tinoco, J., Babcock, R., Hincenberg, I., Medwadowsky, B., Miljanich, P., and Williams, M. A., 1979, Linoleic acid deficiency, *Lipids* **14**:166.

Tiselius, H. G., 1973, Metabolism of tritium labeled pyridoxine and pyridoxine-5'-phosphate in the central nervous system, *J. Neurochem.* **20**:937.

Todd W. R., Elvehjem, C. A., and Hart, E. B., 1934, Zinc in nutrition of the rat, *Am. J. Physiol.* **107**:146.

Tomarelli, R. M., Linden, E., and Bernhart, F. W., 1952, Nutritional quality of milk thermally modified to reduce allergic reaction, *Pediatrics* **9**:89.

Tower, D. B., 1965, Distribution of cerebral fluids and electrolytes *in vivo* and *in vitro*, *Am. J. Physiol.* **208**:682.

Trabucchi, M., Cheney, D. L., Hanin, I., and Costa, E., 1975, Application of principles of steady-state kinetics to the estimation of brain acetylcholine turnover rate: Effect of oxotremorine and physostigmine, *J. Pharmacol. Exp. Ther.* **194**:47.

Trams, E. G., and Brady, R. O., 1960, Cerebroside synthesis in Gaucher's disease, *J. Clin. Invest.* **39**:1549.

Truitt, E. B., Jr., Bell, F. K., and Krantz, J. C., Jr., 1956, Anesthesia. LIII. Effect of alcohols and aldehyde on oxidative phosphorylation in brain, *Q. J. Stud. Alcohol* **17**:594.

Tsuchiya, K., 1969, Causation of "Ouch-Ouch" disease, an introductory review, *Keio. J. Med.* **18**:181.

Terhune, M. G., and Sandstead, H. H., 1972, Decreased RNA polymerase activity in mammalian zinc deficiency, *Science* **177**:68.

Underwood, E. J., 1971, "Trace Elements in Human and Animal Nutrition, 3rd ed., Academic Press, New York.

United States Department of Agriculture, 1979, Agriculture Handbook No. 8-4: Composition of foods: Fats and oils; raw, processed, prepared. U.S. Government Printing Office, Washington, D.C.

United States Department of Commerce, National Research Center, 1976, Report No. NAS/ACT/P-831, Recommendations for Prevention of Lead Poisoning in Children, Springfield, Va.

United States Department of Health, Education and Welfare, DHEW Publ. No. 79-55071, 1979, Surgeon General's Report on Health Promotion and Disease Prevention, U. S. Government Printing Office, Washington, D.C.

Usher, R. H., and McLean, F. H., 1974, Normal fetal growth and the significance of fetal growth retardation, *in* "Scientific Foundations of Pediatrics" (J. A. Davis and J. Dobbing, eds.), Saunders, Philadelphia.

Vergroesen, A. J., and Gottenbos, J. J., 1975, The role of fats in human nutrition: An introduction, *in* "Role of Fats in Human Nutrition" (A. J. Vergroesen, ed.), pp. 1–41, Academic, New York/London.

Verma, K., and King, D. W., 1967, Disorders of the developing nervous system of vitamin E deficient rats, *Acta Anat.* **67**:623.

Victor, M., Altschule, M. D., Holliday, P. D., Goncz, R. M., and County, A., 1957, Carbohydrate metabolism in brain disease. VII. Carbohydrate metabolism in Wernicke's encephalopathy associated with alcoholism, *Arch. Intern. Med.* **99**:28.

Victor, M., Adams, R. D., and Collins, G. H., 1971, The Wernicke-Korsakoff syndrome. A clinical and pathological study of 245 patients, 82 with postmortem examination, *in* "Contemporary Neurology Series," No. 7 (F. Plum and F. H. McDowell, eds.), Davis, Philadelphia.

Volpe, J. J., and Kishimoto, Y., 1972, Fatty acid synthetase of brain. Development, influence of nutritional and hormonal factors and comparison with liver enzyme, *J. Neurochem.* **19**:737.

von Euler, U. S., 1956, "Noradrenaline," Charles C Thomas, Springfield, Ill.

von Euler, U. S., and Gaddum, J. H., 1931, An unidentified depressor substance in certain tissue extracts, *J. Physiol.* **72**:74.

Wald, G., 1968, Molecular basis of visual excitation, *Science* **162**:230.

Wannag, A., 1976, The importance of organ-blood mercury when comparing fetal and maternal rat organ distribution of mercury after methyl mercury exposure, *Acta Pharmacol. Toxicol.* **38**:289.

Warnock, L. G., and Burkhalter, V. J., 1968, Evidence of malfunctioning blood-brain barrier in experimental thiamine deficiency in rats, *J. Nutr.* **94**:256.

Warshaw, A. L., Walker, W. A., and Isselbacher, K. J., 1974, Protein uptake by the intestine: Evidence for absorption of intact macromolecules, *Gastroenterology* **66**:987.

Watanabe, H., and Passonneau, J. V., 1973, Factors affecting the turnover of cerebral glycogen and limit dextrin in vivo, *J. Neurochem.* **20**:1543.

Waxman, S., Corcino, J. J., and Herbert, V., 1970, Drugs, toxins and dietary amino acids affecting vitamin B_{12} or folic acid absorption or utilization, *Am. J. Med.* **48**:599.

Weathersbee, P. S., and Lodge, J. R., 1979, Alcohol, caffeine and nicotine as factors in pregnancy, *Postgrad. Med.* **66**(3):165, 170. *J. Reprod. Med.* **21**:63.

Weber, G., 1969, Inhibition of human brain pyruvate kinase and hexokinase by phenylalanine and phenylpyruvate. Possible relevance to phenylketonuric brain damage, *Proc. Nat. Acad. Sci. USA* **63**:1365.

Weiner, G., 1970, The relationship of birth weight and length of gestation to intellectual development at ages 8 to 10 years, *J. Pediatr.* **76**:694.

Weiss, D., Whitten, B., and Leddy, D., 1972, Lead content of human hair 1871–1971, *Science* **178**:69.

Welbourn, H. G., 1955, The danger period during weaning, *J. Trop. Pediatr.* **11**:34.

Wells, M. A., and Dittmer, J. C., 1967, A comprehensive study of the postnatal changes in the concentration of the lipids of developing brain, *Biochemistry* **10**:3169.

Wheeler, T. G., Benolken, R. M., and Anderson, R. E., 1975, Visual membranes. Specificity of fatty acid precursors for electrical responses to illumination, *Science* **188**:1312.

White, H. B., Galli, C., and Paoletti, R., 1971, Brain recovery from essential fatty acid deficiency in developing rats, *J. Neurochem.* **18**:869.

Whitley, J. R., O'Dell, B. L., and Hogan, A. G., 1951, Effect of diet on maze learning in second-generation rats: Folic acid deficiency, *J. Nutr.* **45**:153.

Widdowson, E. M., and McCance, R. A., 1960, Some effects of accelerated growth. I. General somatic development, *Proc. R. Soc. London Ser. B* **152**:188.

Williams, R. R., 1936, Structure of vitamin B_1, *J. Am. Chem. Soc.* **58**:1063.

Williamson, D. H., and Buckley, B. M., 1973, Role of ketone bodies in brain development, *in* "Inborn Errors of Metabolism" (F. A. Hommes and C. J. Van den Berg, eds.), pp. 81–96, Academic Press, New York.

Williamson, D. H., Bates, M. W., and Krebs, H. A., 1968, Activity and intracellular distribution of enzymes of ketone body metabolism in rat liver, *Biochem. J.* **108**:353.

Williamson, D. H., Bates, M. W., Page, M. A., and Krebs, H. A., 1971, Activities of enzymes involved in acetoacetate utilization in adult mammalian tissues, *Biochem. J.* **121**:41.

Winick, M., 1970, Nutrition and nerve cell growth, *Fed. Proc.* **29**:1510.

Winick, M., 1976, "Malnutrition and Brain Development," Oxford Univ. Press, London.

Winick, M., and Noble, A., 1966, Cellular responses in the rat during malnutrition at various ages, *J. Nutr.* **89**:300.

Winick, M., and Rosso, P., 1969, Effects of severe early malnutrition on cellular growth of human brain *Pediatr. Res.* **3**:181.

Winick, M., Brasel, J. A., and Rosso, P., 1972a, Nutrition and cell growth, *in* "Nutrition and Development," Vol. 1, "Current Concepts in Nutrition" (M. Winick, ed.), Wiley, New York.

Winick, M., Rosso, P., and Brasel, J. A., 1972b, Malnutrition and cellular growth in the brain: Existence of critical periods, *in* "Lipids, Malnutrition and the Developing Brain" (K. Elliott and J. Knight, eds.), pp. 199–212, Elsevier, Amsterdam.

Witten, P. W., and Holman, R. T., 1952, Polyethenoid fatty acid metabolism. VI. Effect of pyridoxine on essential fatty acid conversions, *Arch. Biochem. Biophys.* **41**:266.

Witting, L. A., 1974, Vitamin E-PUFA lipid relationship in diet and tissues, *Am. J. Clin. Nutr.* **27**:952.

Wolbach, S. B., and Hegsted, D. M., 1952, Vitamin A deficiency in the chick, *Arch. Pathol.* **54**:13.

Wolf, G., Korpes, T. C., Masu-Shige, S., Schreiber, J. B., Smith, M. J., and Anderson, R. S., 1979, Recent evidence for participation of vitamin A in glycoprotein synthesis, *Fed. Proc.* **38**:2540.

Woolley, D. W., and White, A. G. C., 1943, Production of thiamine deficiency disease by feeding of pyridine analogues of thiamine, *J. Biol. Chem.* **149**:285.

Wurtman, R. J., 1979a, Precursor control of transmitter synthesis, *in* "Nutrition and the Brain," Vol. 5 (A. Barbeau, J. H. Growdon, and R. J. Wurtman, eds.), pp. 1–12, Raven, New York.

Wurtman, R. J., 1979b, Sources of choline and lecithin in the diet, *in* "Nutrition and the Brain," Vol. 5 (A. Barbeau, J. H. Growdon, and R. J. Wurtman, eds.), pp. 73–81, Raven, New York.

Wurtman, R. J., Rose, C. M., Mathysse, S., Stephenson, G., and Baldessarini, R., 1970, L-Dihydrophenylalanine: Effect on S-adenosylmethionine in brain, *Science* **169**:395.

Wurtman, R. J., Larin, F., Mostafapour, S., and Fernstrom, J. D., 1974, Brain catechol synthesis: Control of brain tyrosine concentration, *Science* **185**:183.

Wurtman, R. J., Hirsch, M. J., and Growdon, J. H., 1977, Lecithin consumption raises serum free choline levels, *Lancet* **II**:68.

Wyatt, R. J., Chase, T. N., Scott, J., and Snyder, F., 1970, Effect of L-DOPA on the sleep of man, *Nature (London)* **228**:999.

Wyatt, R. J., Termini, B. A., and Davis, J., 1971, Biochemical and sleep studies of schizophrenia: A review of literature 1960–1970, *Schizophr. Bull.* **4**:10.

Wyngaarden, J. B., 1970, Genetic control of enzymatic activity in higher organisms, *Biochem. Genet.* **4**:105.

Yahr, M. D., Duvoisin, R. C., Mendoza, M. R., Shear, M. J., and Barrett, R. E., 1972, Modification of L-DOPA therapy of parkinsonism by alpha-methyldopa hydrazine (MK 486), *Trans. Amer. Neurol. Assoc.* **96**:55.

Yosselson, S., 1976, Drugs and Nutrition, *Drug Intell. Clin. Pharm.*, **10**:8.

Yost, K. J., 1979, Some aspects of cadmium flow in the U.S., *Environ. Health Perspect.*, **28**:5.

Yusuf, H. K. M., and Dickerson, J. W. T., 1978, Disialoganglioside G_{Dla} of rat brain subcellular particles during development, *Biochem. J.*, **174**:655.

Zamenhof, S., van Marthens, E., and Margolis, F. L., 1968, DNA (cell number) and protein in neonatal brain: Alteration by maternal dietary protein restriction, *Science* **160**:322.

Zamenhof, S., van Marthens, and Grauel, L., 1972, DNA cell number and protein in rat brain: second generation (F_2) alteration by maternal (F_0) dietary protein restriction, *Nutr. Metab.* **14**:262.

Zamenhof, S., Hall, S. M., Grauel, L., van Marthens, E., and Donahue, M. J., 1974a, Deprivation of amino acids and prenatal brain development in rats, *J. Nutr.* **104**:1002.

Zamenhof, S., van Marthens, E., and Grauel, L., 1974b, Prenatal cerebral development: Effect of restricted diet and reversal by growth hormone, *Science* **174**:954.

Zeman, F. J., 1970, Effect of protein deficiency during gestation on postnatal cellular development in young rat, *J. Nutr.* **100**:530.

Zeman, F. J., and Stanbrough, E. C., 1969, Effect of maternal protein deficiency on cellular development of fetal rat, *J. Nutr.* **99**:274.

Zeman, W., Donahue, S., Dyken, P., and Green, J., 1970, The neuronal ceroid-lipofuscinoses (Batten-Vogt syndrome) *in* "Handbook of Clinical Neurology," Vol. 10, (P. J. Vinken, ed.) p. 212, North-Holland, Amsterdam.

Zetter, G., 1970, Biologically active peptides (substance P), *in* "Handbook of Neurochemistry," Vol. 4 (A. Lajtha, ed.), pp. 135–148, Plenum, New York.

Zivin, J. A., and Snarr, J. F., 1972, D(-)-3-hydroxybutyrate uptake by isolated perfused rat brain, *J. Appl. Physiol.* **32**:1664.

INDEX

INDEX